Tūpuna Rock

Tūpuna Rock

*Lost, stranded and assumed dead,
the ancestors' spirits must guide them.*

Klaus Brauer

Tūpuna Rock
a *Limestone Point* book

Copyright © 2024 Klaus Brauer
ISBN: 979-8-9897195-0-1

All rights reserved. No part of this book may be reproduced or transmitted in any form electronic or mechanical without the written permission of the author, except for the inclusion of brief quotations in a review.

The author hereby grants permission for the reproduction of his drawings and associated text from Chapters 4, 5, 6, 7 and 8 in support of science and mathematics education.

Text and drawings by Klaus Brauer

The chart in chapter 3 is based on Land Information New Zealand (LINZ) data which are licensed by **LINZ** for re-use under the **Creative Commons Attribution 4.0 International** license.

Cover art by Richard Fraker

To the Reader

Enjoy the story. You can read without slowing to study any of the science or mathematics described in this book. On the other hand, you may enjoy taking in all the detail – at first read or by returning later.

What matters, is that the novel's characters came to understand and use the knowledge and skills of the ancient voyagers – as their extented Pacific clan, inheritors of their ancestors' genius, can today.

Contents

1	Fishing	1
2	Afloat	10
3	Ashore	26
4	Latitude	39
5	Longitude	57
6	Tūpuna	69
7	Toki	91
8	Making Ready	113
9	Sea Trial	137
10	Voyagers	145
11	Intercepted	159

Author's note	179
Acknowledgements	180
Glossary	181
Star Names	183
Celestial latitude at the time of the migration	184
Bibliography	185

mō ngā tamariki ō Te Moana-nui-a-Kiwa

for the children of the Pacific

1 Fishing

Under a clear blue sky, the rocky ridge rising from the northern shore of Whatupuke Island was awash in sunshine. The ridge, in turn, radiated warmth complementing that of the direct midday sun. It was the day after Christmas, summer there in the Southern Hemisphere. Barefoot, wearing only a floppy hat, t-shirt, shorts and life jacket, the young skipper leaned back in the cockpit and let the boat drift for a minute as he thought about his classmates shivering back home in Vancouver.

The steep, 200-metre-tall ridge sheltered the boat from the wind that had provided a sporty sail out to the islands. While his older sister, Anna, enjoyed fishing in the calm water, Per, the skipper, played with the boat, sailing the ten-metre sloop slowly back and forth. He occasionally dropped a float over the side and, after sailing onward for a bit, practiced a man overboard recovery manoeuvre, coming about to sail slowly back to the float and bringing the boat to a halt a metre to windward of the float. As the boat drifted down on the float, he 'rescued' it with the boathook.

The fishing was less successful than the man overboard drills, but no one was keeping score. Fishing was therapeutic for Anna, just as sailing was for Per. Both activities require continuous attention; but the required level of attention is low, aside from when there is a fish on the line or the sailing conditions turn challenging. That sustained, low level of attention is ideal for pushing day to day worries to the side. Per, generally a good student, was struggling in French, the least favourite subject for many western Canadians, but French was absent from his thoughts as he sailed back and forth in the calm water downwind of Whatupuke. Anna was in her final year of high school, in the midst of applying for university admission. Some university applications were due within weeks, but under the bright blue sky that afternoon, she was focused on the lures she played in the water, without a thought of applications or essays.

As the afternoon wore on, the other boats in the area departed one by one and headed back to port. A man at the helm of one of the last to leave gave a dramatic wave to the teens, sweeping his arm toward the west, toward the harbour entrance twenty kilometres away.

'What do Kiwis do on Boxing Day?' Per wondered aloud.

'Eat Christmas leftovers, I think.'

'That dude must have a lot of leftovers he needs help with.' With that quip, Per dismissed the gesture.

In time, the drama of the departing boater's wave began to trouble Per. He had been watching the wind waves rolling through the gap between Whatupuke and Motu Muka, the next island to the west. The wave heights hadn't suggested strong winds, but closer examination revealed the tops being blown off waves. 'I think it's time to head in,' he announced as he assessed the sails. He had the midsize jib up and the first reef in the mainsail, appropriate to his estimate of the wind driving the waves. 'You know what?' he added almost as an afterthought. 'Just to be safe, we should have our harnesses on.' They donned their harnesses and clipped the tethers to the jackline running along the cockpit floor.

As soon as they emerged from Whatupuke's wind shadow, a gust knocked the sloop down nearly flat. 'Let out the jib!' Per yelled to Anna above the howling wind as he eased the mainsail out. He started the engine and ran it up to full power to get into Motu Muka's wind shadow as quickly as possible.

With the sloop back on her feet, Per looked to the south, through the gap between Whatupuke and Motu Muka, saying no more than, 'Holy crap!' Whatupuke's high ground, which had sheltered them from the wind, had also blocked the view to the south, where mountainous black storm clouds had formed. Massive wind-blown waves were breaking in the shallows and rocks between the islands, leaving only smaller, newly formed wind waves to emerge from the gap. Those smaller waves had deceived Per's estimate of the wind.

Once snuggled into the lee of Motu Muka, Anna took the tiller to steer the sloop's bow into the diminished wind while Per went forward, hauled down the jib and tucked the second reef into the mainsail. After stowing the midsize jib below, he emerged with the storm jib, a small bright-orange headsail, which he raised from the bow.

Returning to the cockpit, Per studied the churning surf in the twenty kilometres of open water between Motu Muka and the mainland. He went below once more to retrieve wet-weather jackets and trousers. 'Yeah, it's not raining,' he hesitated before adding, 'yet. But waves are going to be bashing us on the windward beam all the way. It'll be a pretty wet ride, even if we get back before any rain comes.'

While donning his wet weather gear, Per began to explain the plan to Anna. 'We've got to make due west to clear Bream Head. I'd rather make a bit south of west to properly enter the fairway into port.

'The engine is great for motoring this boat along when there's no wind, but the engine itself isn't powerful enough to push through the wind and waves I see out there. We'll need sails up, too.

'The storm jib, flying from the bow, will want to turn the boat toward the north, so stand by to ease it out if I ask. The main will have the opposite effect, wanting to turn the boat to the south. I'll manage the main. We'll be managing our heading with both the sails and, of course, the engine streaming water past the rudder. Make sense?'

'Yup.'

Anna was not an experienced sailor, but he knew she could figure it out. It was all just forces and angles, and she could do trigonometry in her sleep. Per took the helm while Anna put on her wet weather gear. He checked that their life jackets, harnesses and tethers were all secure before asking, 'Ready?'

'Yup.'

'Here we go.' Per turned the bow to the west and pushed the throttle forward.

Once clear of Motu Muka's shelter, still somewhat sheltered by a few smaller islets, waves began pounding on the little sloop's windward bow, and it began to turn toward the north, away from their heading back to safety. Anna let out the jib enough to keep them headed slightly to the south of Bream Head. Waves beating against the windward side of the hull threw spray by the bucketful over the rail into the cockpit. As expected, it had become a wet ride.

Anna was able to steer with the storm jib while the smaller islets provided some shelter, but once the sloop emerged from the islets' shelter, the mainsail was overpowered by the waves beating against the windward bow, and the boat began to turn toward the north. 'Just let the jib run out,' Per called over the howling wind as he hauled the mainsail in hard to turn the boat back to the west.

The little sloop was losing speed and the rudder becoming less effective when a huge gust hit. With a low boom like the bursting of a massive balloon, the mainsail ripped from its aft edge all the way forward to the mast. Out of control, the bow spun northward as the useless mainsail flapped angrily.

'So much for clearing Bream Head.' Preoccupied with keeping them afloat, Per was emotionless.

'Where then?' Anna asked apprehensively.

'We run for Ngunguru Bay, find shelter behind the head by Pataua at the south end of the bay.' Looking north, they could make out the head in the distance.

Per caught Anna glancing up at the torn mainsail flapping uselessly in the wind. 'Grandpa Koro won't give a rat's ass about the sail, if we get ashore safely,' he assured.

'Per, I'm gonna send mom, dad, Grandpa Koro and Grandma Karani all a text to tell them what we're doing.'

'Good idea.'

'I turned my phone off so it wouldn't be searching while we were out here. It's down below in my dry bag. Can I use yours?'

'Yeah, that's probably better. You send texts all the time. I don't know about mom and dad, but you send so many texts that I don't look at half of them.'

Per's attempt at humour fell flat. 'So my texts are spam?' Anna asked in a hurt tone.

'No, I just meant, since I'm usually too lame to let them know what I'm up to, they're sure to open a text from my phone.'

Anna tapped out the message on Per's phone and hit 'send'. After a minute she reported, 'It didn't go. No reception. I'll check for reception every now and then, and resend when we have some.'

Per didn't hear Anna's last sentences; he was focused on the boat. 'Annie,' he commanded, 'winch in the storm jib a bit. Manage it to pull us northward.'

Per's mind was consumed with sailing the boat with a new wave overtaking them every few seconds, surfing down the face of one wave before falling into a trough and being beaten on the stern by another. Some of the thumping on the stern sounded leaden, and Per turned to find the small dinghy they had been towing, now swamped, was being thrown against the stern by the advancing waves.

In an instant the tiller controlling the rudder became stiff, intermittently pulling hard to starboard. Per looked astern to discover that the tiller pulled just when the dinghy had fallen back to the full length of its tether. As he fought the tiller, he shouted over the noise of the wind and crashing waves, 'Annie, I need your help back here.'

Anna secured the jib sheet and joined Per at the stern.

'The dinghy's tether is fouled on the rudder. I can't steer. I need you to cut the tether.'

Anna gave an inquiring look.

'Yeah,' Per offered, 'Grandpa Koro won't give a rat's ass about the dinghy either.' With a free hand he gave Anna his rigging knife.

Watching Anna lean out over the stern of the pitching boat, trying to seize an opportunity to cut the tether as the dinghy fell back repeatedly before being thrown forward against the stern again and again, Per suddenly called out, 'No! The dinghy will break your arm before you have a chance to cut the tether. Just cut the end secured to the sloop; maybe a loose end in the water will allow the tether to work itself free.'

Shortly, when the dinghy snapped once again to the end of its tether, there was a mean crunching sound from inside the hull, and the dinghy disappeared behind a wave. The tiller moved freely, too freely, and had no apparent effect on the rudder. Per threw open the hinged bench on the port side of the cockpit and leaned in to inspect the connection to the rudder. The rudder was completely gone, and water was flooding into the boat through the hole left by the rudder stock.

Per jumped down past the hinged bench, landing on the bottom of the hull with a splash. Lying in the water, flat on his chest, he reached under the cockpit floor to cover the hole left by the rudder with his palm while searching for something to plug the hole. Turning his head upward he called to his sister, 'Look for something to plug a hole about five centimetres across.' In a few moments, Anna appeared above him holding the conical wooden plugs kept to plug broken pipes. 'Yeah! Give me the biggest one.' He took the plug but handed it back after a moment, 'It's too small and the hole's ragged.'

Anna returned with the plug wrapped in a towel. 'Is that big enough?'

'Yeah, I think so.'

The boat pitched and rolled as Per worked the plug into the hole. Seawater sloshing around filled his nose and left him sputtering several times, but he finally got the plug firmly in place and jammed a tool box between the top of the plug and the underside of the cockpit floor to hold the plug down.

Per was back in the cockpit, studying the boat's motion in the waves overtaking it, when the water down below found its way to the engine's air intake. The engine didn't sputter; it stopped dead, seized by a cylinder full of water. It was finished.

The storm jib remained intact, alternately pulling the boat along and collapsing as the stern sashayed in the waves. Needing some sort of rudder to stabilize the boat's motion and keep the storm jib inflated, Per recalled a trick he had once read. Retrieving the backup anchor from its nearby stowage, he secured the anchor to a cleat on the port side at the end of ten metres of line and then eased it overboard. The drag of the anchor kept the boat turned just enough to port to keep the jib inflated on the starboard side.

As Anna threw bucket after bucket of water out of the cabin, Per evaluated the situation. It was definitely time to call for help. He glanced at the GPS, pulled the hand-held VHF radio from a pocket of his life jacket, pressed the transmit button and spoke slowly and deliberately. 'Mayday. Mayday. Mayday. This is Manaia, Manaia, Manaia. Location about four nautical miles northeast of Bream Head. Have lost rudder and engine. Taking on water. Running northward under storm jib. Two souls onboard. Mayday, Manaia. Manaia out.'

After releasing the transmit button, Per listened eagerly for a reply. After thirty seconds he heard a weak broadcast, perhaps a reply, rendered undecipherable by static and the howling wind. He was puzzled by the weakness of the reply, having expected a reply to boom in from a high-powered Navy or Coastguard ground station. After waiting another two minutes, he repeated his mayday call, with the same result.

The weather continued to deteriorate. Heavy rain arrived and restricted visibility, obscuring even the nearby coast. Waves continued to pound the stern, often crashing over the transom, rolling through the cockpit and tumbling down into the cabin. Anna's bailing began to seem fruitless. 'Per?' she called up to into the cockpit, 'The galley sink drains overboard, doesn't it?'

'Yeah. Slowly, but yeah.'

'How about if I put the hatch boards in to stop the waves from washing into the cabin? I could bail into the sink.'

'Yeah, good idea.'

The hatch boards covered the full height of the entry to the cabin leaving only a sliding hatch cover that extended horizontally from the cabin roof. Anna and Per would not be able to see one another as she bailed and he managed the boat from the cockpit. That thought troubled her. 'You gonna be okay out there?'

'Yeah. I can't get much wetter. You okay?'

'Uh huh,' she replied unconvincingly.

'Just think about bailing.'

~ ~ ~

Shifting the jib from one side of the boat to the other, while moving the dragged anchor to the opposite side, Per convinced himself he had some limited ability to steer the boat. With no visibility and daylight fading, hoping to outlast the storm, he chose to just follow the coast northward as best he could, guided by the GPS clipped to a post standing in the centre of the cockpit. The ship's compass mounted at the top of the post was bouncing around so wildly in the pitching boat that it wouldn't settle on a bearing. Per pulled his hand-bearing compass from a pocket of his life jacket, looped its lanyard around his neck and held it as level as possible to get compass bearings.

As the daylight filtering into the cabin through the overhead hatch faded, Anna switched on the red cabin light – red so any light spilling out wouldn't compromise Per's night vision. The pitching and rolling of the boat made half of the water she poured into the sink tumble out again onto the cabin floor, but she was making some progress, and the activity kept her mind occupied.

Anna heard Per suddenly scream, 'Hold on!' as his legs appeared, descending through the open hatch. He dropped to the cabin floor and slammed the hatch cover shut. In a moment the sloop was on her side. A wire rope shroud that held the upper half of the mast straight, failed. Drowning out the sound of pots, pans and everything loose in the cabin tumbling around, the whiny shriek of the aluminium mast folding in half was telegraphed through its metal to the cabin roof, in which it resonated like the soundboard of some musical instrument from hell.

As the boat continued to roll, Anna and Per found themselves sitting on the ceiling of the inverted cabin in water up to their waists. More rigging failed, and the mast was ripped from the cabin roof; the water then covering the roof muffling the sound to an ugly thud. The cabin light continued to burn on the ceiling of the overturned cabin. Covered by rippling water, its red light was refracted into sparkles and flashes around the cabin, and in the stunned faces of the crew.

2 Afloat

The patch of blue sky took her by surprise. For days and nights, in a sea whipped by gale-force winds, the whole world had been nothing but shades of grey and black. By day, the sky had been a turbulent ashen colour and the sea an angry scene of slate streaked and dotted by the grey foam of breaking waves. By night, there was nothing but blackness, and the punches thrown by the sea and sky arrived without warning.

The weather had begun to settle down. The wind was no longer a full gale, but remained strong, constantly rattling and ruffling loose bits of Manaia's destroyed rig and occasionally blowing spray over the cockpit. Covered by low clouds, Per and Anna had been denied a sense of direction as the sloop wallowed in the advancing wind-driven waves.

It was their fourth day adrift, when the hole opened up in towering clouds, and Anna found herself staring up through the clouds, as if looking up from the bottom of a rock-lined well, toward a small patch of blue sky. She studied the motion of the boat as it skidded backwards down the waves overtaking it, turned on her phone for a moment to glance at the time and then studied the sunlight playing on the edge of the hole toward the advancing waves.

'Per,' she called down into the cabin, 'The waves are pushing us northward.'

Anna's younger brother poked his head out of the cabin, 'How can you tell?'

Tūpuna Rock

'I can see a bit of blue sky through a hole in the clouds. The edge of the hole the waves are pushing us away from is lit by the sun. That means we're being pushed toward the sun. It's about noon, so here in the southern hemisphere that's northward.'

Per turned his head upward, squinting for the first time in many days. 'Brilliant,' he observed of his sister, and perhaps of the blue sky as well.

'How are you feeling?' Anna asked.

'Better. I got some sleep and haven't puked since yesterday. I'm so used to how a boat should feel under sail that, when the motion is all wrong, my whole body rebels.'

'All wrong? You mean rolling over and tumbling down monster waves for days on end is wrong?' Anna forced a smile meant to lift her brother's spirits a bit.

'Yup. Boats aren't supposed to do rolls. Still, it's kind of funny that your sailor bro was the seasick one.'

'Not funny. It's been horrible. And, even when you were puking, you got everything that was still with the boat tied down.'

'Still, it is funny that my nerdy, non-sailor sister was keeping her lunch down while I was heaving mine over the side.'

'Maybe I'm becoming a sailor.'

'But still a nerd.'

'Always a proud nerd.'

'Okay, proud nerd, are those clouds telling you we're headed due north, or maybe northeast or northwest?'

As Per bailed a few bucketsful of water out of the cabin, Anna looked up and studied the clouds for a full minute, looking for some surface that would tell her the answer. 'Sorry, "northward" is the best I can do. Definitely not southward. And not due eastward or due westward, but somewhere in between, vaguely toward the north.'

'Well, since we lost our GPS and compasses when we were rolled, that's the best we've got, and it is something. It's a big something.' After some thought Per added, 'Jeez, I wish I had the sólarsteinn Farfar brought me from Bergen. You know, that crystal Vikings used to find the direction to a sun hidden behind clouds.'

'Have you tried it? Does it work?'

'I tried it a few times on shore, looking between trees – not ideal conditions – but I could see something. I think it kinda worked. With practice I think I could use it. The Vikings did find their way across one thousand kilometres of cloud-covered ocean to Iceland.' Finally he acknowledged ruefully, 'But I don't have it; it's in my desk drawer at home.'

A whisper of a voice crackled from the handheld radio clipped to Per's life jacket. The voice was broken and faint, but definitely a human voice. Per climbed up out of the cabin into the cockpit, waited thirty seconds after the voice stopped, stood, raised the little radio, pressed the transmit button and spoke slowly, 'Mayday. Mayday. Mayday. This is Manaia, Manaia, Manaia. Mayday, Manaia.' He hesitated and then concluded, 'Two souls on board. Manaia out.'

Anna gave him an inquiring look. She had heard enough mayday declarations in the previous days to know he was supposed to include their location and the nature of their distress.

'Yeah,' Per acknowledged, 'I don't have a clue where we are, and it would take a lot of transmit time to fully explain our situation – taking on water, no rudder, no engine, no mast, no navigation equipment, running out of food… I think it's better to save battery power just so we can put out a call someone might hear. If anyone out there knows we're still afloat, they'll keep searching for us.'

'We are still afloat,' Anna noted.

'Yeah, but we may be the only people who know that.'

'We are still afloat,' Anna repeated, intoning it as best she could to be reassuring.

'Yup, we are still afloat, as long as we keep bailing faster than we take on water.'

Per clipped his harness tether to the jackline running from cockpit to bow and silently started forward. He crept cautiously along the narrow deck on the port side between the cabin and the deck's edge. Crawling over the broken mast, he slid one hand along the grab rail on the cabin roof and then the other hand along the lifeline supported above the deck's edge by pipe-like stanchions. The stainless-steel-wire lifeline, the only thing standing between Per and the surging sea, was unreassuringly slack. When he gave the lifeline a tug, it moved easily, pulling the nearest stanchions with it. Glancing over the cabin top he confirmed his recollection of the starboard-side lifeline. The line itself was gone. Some of stanchions had been ripped away, those that remained were badly mangled.

The scene on the bow was more orderly, jury-rigged but orderly. The loop of nylon line he had rigged to cushion the tugs of the anchor chain run out over the bow roller remained sound. He studied the boat's motion, satisfied that the drag of the anchor and twenty metres of chain paid out into the sea from the boat's bow was stabilizing its backward, stern-first motion.

Looking back along the port side Per studied the broken mast he had secured there while vomiting over the side, satisfied that it was still secure. Looking back along the starboard side, he studied the boom he had secured to the mangled stanchions. The mainsail, still attached to the boom, was fluttering in the wind. He made his way along the starboard side of the surging deck and secured the mainsail to stop its fluttering with his left hand while holding onto the cabin-top grab rail for dear life with his right hand.

Convinced that everything of consequence that remained onboard was secure, Per crawled backward to the bow and freed the shredded storm jib, tucking the flapping bits of orange polyester under his life jacket. He finally made his way back to the cockpit, tossed the bits of the storm jib into the cabin and sat with his back against the cabin bulkhead.

Thinking about the wreckage he commanded, Per studied the stump of the post that had once supported a GPS receiver and the ship's compass. He raised his hand to feel the rope burn on his neck and the bandage on his jaw.

'Do you remember how that happened?' Anna asked gently.

'My hand-bearing compass was on a lanyard around my neck. When I jumped down into the cabin, it must have snagged on something. I'm glad the lanyard broke before my neck did, but I sure wish we had that compass.'

'It sawed quite a gash in your jaw. Does it hurt a lot?'

'I think it does. But my brain's in some other place. I can feel my jaw. It feels like pain, but there's so much more going on in my head that it doesn't register somehow,' as an afterthought he added, 'if that makes any sense.'

Per and Anna were lost in their thoughts for minutes, each with their own perspective on all the things going on in the head of the fifteen-year-old skipper. Finally, Anna asked, 'Per, the water in the cabin was already nearly knee deep when we were rolled. If we'd rolled with the hatch open, that would have been it. What made you suddenly jump down into the cabin and slam the hatch shut? I didn't hear or feel anything different from the churning that had been going on for hours. I didn't know anything was happening until you screamed, "Hold on". Did you see something? Hear something? Feel the boat start to roll?'

'It was a new moon, too dark to see squat,' he began, shaking his head. 'I didn't hear anything different, and the boat hadn't started to roll, but I could feel something through the hull. I could feel a strange, low vibration, something I've never felt before. The sea was angry. It was groaning, in a pitch too low to hear, but I could feel it through the hull.'

'With your bony butt?' Anna asked with a kindly smile.

'Yup, with my sensitive, bony butt,' Per answered returning the smile.

Being so different, Per guided by intuition and Anna guided by calculation, the siblings had grown up with little cause to compete with or feel threatened by each other. Rather, each was by turns amused and impressed by the other's forte. The playful 'math nerd' and 'sensitive, bony butt' jabs were expressions of their appreciation of the other's strength. That the jabs also provided some amusement was a blessing. In their dire situation, they were just barely holding themselves together emotionally. Every smile helped.

Per looked across the cockpit at his sister, 'The truth is I wasn't throwing up just because of the boat's motion. I was scared, really scared. I'm still really scared, just not quite so much.'

'You sure didn't panic and curl up into a ball on the cockpit floor. You plugged the hole where the rudder was ripped out. You secured the pieces of mast and sails. You hung the anchor off the bow to turn us into the waves, so the cockpit wasn't flooded by the waves overtaking us. You kept functioning.'

'Do you remember what Grandpa Koro tells me every time I take the helm? "Keep everyone safe." He always says the most important job of a skipper is to keep everyone safe. As I was retching over the side, holding onto the boat with one hand and tying things down with the other hand, I wasn't thinking about the knots I was tying; I was just saying over and over in my head, "Keep everyone safe. Keep everyone safe."'

'I'm sorry I asked you take me fishing out by the islands,' Anna announced, voicing guilt she had been processing for days. 'We could have had a nice day sailing around the bays close to the marina.'

'Annie, I'm the skipper. It was my decision to go. And I didn't pay enough attention to the weather.'

'Can we say we're even?'

'Okay, we're even,' Per allowed with a laugh.

'And we're still afloat.'

'Still afloat, somewhere in the South Pacific, or the Tasman Sea.'

'Could we actually be in the Tasman, to the west of New Zealand?'

'I tried to outrun the storm. We were nearly at the northern tip of New Zealand when we lost everything. The sea's been confused. With the wind and waves churning as they have been, we could have been pushed in any direction from there. Yeah, we could be to the west of New Zealand.'

'So even if we get an accurate sense of which way is which, unless we see land, we don't know which way to go.'

'Yup, we're lost.'

'…at sea,' Anna completed the thought. 'I've been thinking about it. By now the news is reading, "Canadian teens missing and feared lost at sea."'

Per had also wondered what was thought about them back in New Zealand. He wondered why he hadn't heard more ships and aircraft on the radio searching for them and wondered if they had, in fact, been blown into the Tasman Sea toward Australia.

'Why Canadian teens?' he asked. 'Wouldn't they mention that we're half Norwegian or half Māori?'

'I'm not sure why I said it that way. We've grown up in Canada, a gift from the Commonwealth blending with the Scandinavian diaspora in the great white north. We've just come to New Zealand during school vacations. Maybe Viking or Māori sailors wouldn't get lost? We're complicated, Annie. Maybe it was just that I thought kids from overseas make a better story. I don't know.'

Anna paused in thought for a while, finally asking softly, 'What are mom and dad feeling, and Grandma Karani and Grandpa Koro back in Whangārei?'

Per turned to his sister with a pained look, 'Annie, it's too much to think about. I can't think about it now. I'm the skipper. I've got to get us someplace safe.'

Anna understood.

~ ~ ~

'Do you hear that?' Anna asked. 'Sounds like waves crashing.'

Per paused as he lifted his head and listened, 'Yes. But I can't tell where it's coming from.' The clouds had surrounded them once again, and the sound was echoing around between the clouds. With no more than two- or three-hundred metres of visibility, they wondered in which direction safety, or danger, lay.

Tūpuna Rock

Eventually they could make out that they were drifting past a rock pinnacle rising up out of the sea.

Per studied the pinnacle. Waves were crashing against its vertical walls, promising to wreck any vessel trying to land. 'We haven't seen any land for days. But, if I could steer this boat, I wouldn't steer toward that.'

Per thought about his inability to steer the boat, being totally at the mercy of the wind and waves. The weather had settled down enough that he could finally entertain the idea of sailing what remained of the boat. In time he hatched a plan: Build a short, makeshift mast by rigging the bent mast as an inverted V stood up in the centre of the boat. Hang the boom horizontally from the apex of the inverted V. Hang what remained of the mainsail, which was still attached to the full length of the boom, down from the boom, leaving a more-or-less square rig. Anna offered a pen and notebook from her dry bag, and Per quickly sketched the idea.

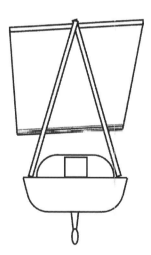

They worked together, lashing a block to the apex of the bent mast and rigging lines fore and aft from the apex to hold the jury-rigged mast upright. A steering oar came together from the spinnaker pole and a shelf pulled from the galley. With Anna at the improvised steering oar, Per hauled in the anchor using a winch handle to turn the capstan. As the anchor came aboard and Anna held the steering oar dead astern, the boat slowly turned around to move bow-first in the waves. Finally, Per raised the inverted mainsail to the block at the apex of the inverted V mast, and they began to make way. They were only making two or three knots, but it was something, and the boat's motion was more comfortable.

'Farfar would be happy to see you raising a square rig, a Viking sail,' Anna noted, referring to their dad's father, a Norwegian immigrant to Canada.

Per smiled, 'He would, wouldn't he?'

'Where are we going?' Anna asked.

'With the wind. I think it's the fastest point of sail with a square rig. I don't know where we're headed, but the sooner we get there the better.'

'Have you ever sailed a square rig?'

'Nope. There's a first time for everything.'

'How about your hero, King Olav the fifth?'

'He must have sailed some square rigs. I'm sure there were some replica Viking longships around Norway in his time. Anyone with that kind of boat would invite their king with an Olympic goal medal in sailing to give it a sail. I'm sure he couldn't resist the offer.'

'Skipper, what can I do?'

'We've about finished the canned food Koro had onboard. I'm hungry; you must be, too. Please catch us a fish. And can you put that math brain of yours to work to figure out where in the world we are. I know you don't have much to work with, but anything...'

While playing a variety of fishing lures over the side, Anna looked for any clue that might reveal where they were. Hours passed with no success on either front.

Anna was aware that her favourite lures might not be appropriate for the unfamiliar deep water, so she tried everything in her tackle box. Finally, shortly before nightfall, she reeled in a beautiful haku. Remarkably, the boat's barbeque grill and three full propane canisters were still in the lazarette. The grill was dented from tumbling around as the boat was rolled, but it still worked, even its piezoelectric igniter.

As Anna handed the first piece of fish to Per, he reached out and took her free hand. Holding her hand, he began to sing a prayer their family always sang before meals, 'O …'

Anna joined in, '… Du som metter liten fugl velsign vår mat o Gud.' The song was a settling bit of normalcy, and they enjoyed a modest feast washed down with fresh water from one of the jerry cans full of water their grandfather kept onboard.

The fishing had been a success, but Anna hadn't found a single clue as to where they were.

~ ~ ~

As dawn broke the following morning, they had five to ten kilometres of visibility under a low cloud cover that frustrated their ability to see the sun and get any sense of direction.

'Something's not right,' Per announced. 'There's something odd about the way the boat's rocking, there's kind of a hiccup. Do you feel it?'

Anna held still for a moment, finally admitting, 'Sorry, no.'

'It's kinda subtle. Something heavy is moving and bumping into something else.' Per stood to study each joint in the jury-rigged mast and boom arrangement, 'That's not it.' Neither was it anything on deck. 'I don't like it. Manaia can't stand any more damage. It could have been going on for days. I just couldn't feel it in the rougher seas.' Per gave Anna the steering oar and went down into the cabin. Anna could hear him shuffling things around and finally, after a minute she heard him say, 'Damn!' in a low groan.

'What, Per?'

'Don't move. Stay where you are. I need to figure out what's happening.' Anna could see Per moving slowly from side to side in the cabin, looking down into the flooded bilge where he had removed the floorboards. He finally emerged from the cabin staying as close as possible to the port side, the side where Anna was seated, and sat beside her.

'The bolts that hold the keel onto the boat are pulled partway through the bottom of the hull. The hull under the bolts must have been crushed when we were rolled. That loosened the joint between the hull and the keel. The keel's been swinging back and forth, pulling the bolts farther and farther into, partly through, the hull.'

Anna was familiar with the keel. She'd studied it while snorkelling around Manaia in the marina. Designed to offset the wind's force on the sails as well as the weight of the mast towering overhead, the keel was a massive lead fin that hung down a metre and half from the bottom of the hull. It was no more than twenty centimetres wide where it met the hull. In the way her brain worked, she recalled its shape, quickly estimating its weight as several hundred kilograms with a centre of gravity about a metre below the hull. The narrow attachment coupled with the lever arm of the low centre of gravity multiplied the force on the bolts holding it in place by ten times or more. Anna knew that, in the intuitive way his brain worked, Per understood it all. She did not need to speak of mass, centre of gravity or lever arm.

'We should stay on one side of the boat to minimize the rocking and additional damage,' Per proposed. 'We're both right-handed, so it's more natural to hold the steering oar in our right hands. I vote for staying on the port side.'

'Makes sense.' Anna paused before asking, 'Do you know how thick the hull is where the keel's attached? How much thickness is left holding the keel on?'

'I don't know. She's a well-designed boat, but she's already given us much more than she was designed for.

'If all of that lead pulls the bolts the rest of the way through the hull, the keel will take off like a rocket toward the bottom of the ocean,' Per explained. 'The keel is most of what keeps the boat right side up. If we lose it, water will flood through the bolt holes, but the more immediate danger is that we capsize.'

Per went silent.

'You okay, bro?'

'Yeah, I just need to make a plan.' He thought for a minute. 'If we lose the keel, we've got to drop the mast, the whole rig, as soon as possible, its weight aloft and the wind's force on the sail will lead to a capsize. If we get the rig down quickly enough, we may be able to keep the boat right side up. Once the mast and sail are down, I'll go into the cabin and plug the bolt holes. Then we bail like crazy.'

Per rigged a quick-release mooring hitch in the line from the stern that was holding the mast up against the wind's force on the sail. 'If we lose the keel, we'll know it. The boat will bounce upward as it loses the keel's weight, and there will probably be a god-awful sound.' Holding out the loose end of the mooring hitch, he explained, 'If that happens, pull this as hard as you can. That should let the mast, sail and everything fall down forward. Whoever is at the steering oar should have my rigging knife handy. If the line doesn't release right away or gets tangled, cut it, don't waste time trying to untangle it.'

'Understood,' Anna said as she caught Per's eye. He was wound up tight, anxiety written all over his face. 'I didn't feel anything odd in the way the boat was rocking. That's another time your sensitive butt saved us.'

'We're not saved yet.'

'You sensed a problem before it sank us and made a plan to keep us afloat. What did that take you, five minutes? I call that a save.'

Per took in Anna's appreciation with a smile. 'You didn't mention that my sensitive butt was bony.'

'Yeah, it was your bony sensitive butt that saved us again.'

~ ~ ~

Sometime near midday a pair of rocks came into view. Relieved to see a place to get ashore before the sloop lost its keel, Per managed the steering oar to set a course for the twin rocks. As they closed in on the rocks, Per and Anna studied every detail, looking for a place to land along their sheer walls. Hoping there was no reef between the rocks, Per committed to a course between them, offering two coasts to choose a landing spot from.

A vision of the keel bolts working their way through the bottom of the hull filled his mind as he chose to sail close to the rock on the port side, tempted in desperation to crash Manaia at the foot of a rocky cliff, hoping that he and Anna could scramble to safety before the sea grabbed them and pulled them under. Per unclipped his harness tether from the jackline and told Anna to do the same. No longer secured to the boat, it was clear that he was preparing to abandon ship. He had serious doubts about their chances of getting ashore safely, but knew their chances of survival were better there than if they lost the keel in the open ocean.

'Wait!' Anna shouted suddenly, 'There's another rock up ahead.' Barely visible, kilometres ahead in the distance, there was, indeed, another rock. 'Maybe that one is better.'

'It couldn't be worse.' The vision of the keel bolts was still in his mind, but it was less immediate than a vision of his sister's lanky body being bashed by waves against the rocks. Per closed his eyes and shook his head in relief.

As they sailed safely away from the twin rocks, Anna reminded, 'There is more fish from yesterday. Want a bite?'

'Not yet. I need to settle down for a while.'

~ ~ ~

A silence was broken by a faint, static voice on the VHF. Per waited and then stood slowly in the cockpit, pressed the transmit button on his handheld radio and spoke slowly and deliberately, 'Mayday. Mayday. Mayday. This Manaia, Manaia, Manaia.

Tūpuna Rock

Mayday, Manaia. Two souls on board. Manaia out.' He released the transmit button. The slow pace of his call was good radio discipline, but his tone revealed that he knew the call was hopeless. No one would hear it. He caught Anna's eye, 'They're broadcasting at fifteen watts or more from a tall antenna. I can only put out five watts, two-and-a-half metres above the water,' he looked at the radio and added, 'and, after all the mayday calls I've made, the battery is dying.'

'We are still afloat, Per, and land is in sight.'

The previous encounters with the pinnacle and the twin rocks were not encouraging, but having a potential destination gave Per a renewed sense of purpose. Realizing they could not reach the rock before nightfall; he lowered the sail to let Manaia drift with the wind and waves until daybreak.

~ ~ ~

Per was at the helm as the sky grew bright with morning. The rock was still downwind, perhaps fifteen or twenty kilometres away. It seemed a good time for Anna to take the helm, so he woke her, 'Annie, when is the next high tide?'

Anna shuffled through the papers from her dry bag, 'At 12:52 – back in Whangārei – here, I don't know.'

Per gave the helm to Anna and raised the sail before retiring to the cabin to bail and finally to rest. 'We've got some hours, just steer toward the rock. If I actually fall asleep, wake me before we hit the rock.'

He did not actually sleep, but even a few hours' rest was rejuvenating. After bailing some more, he emerged from the cabin in late morning to study the rock in the distance. 'What time is it?' Per's phone had died days earlier, trying but still failing to send the text.

'I've been keeping my phone off to save battery. Should I turn it on to check?'

'Yeah, it's important. I want to get there at high tide, to have our best chance of getting over any reefs, and to get the boat as high on the shoreline as I can.'

Anna turned on her phone, checked the time and then promptly turned it off again. 'It's two hours and twelve minutes to high tide,' she paused before adding, 'in Whangārei.'

'I know. We're not in Whangārei anymore, Toto. Still, it's the best tide estimate we've got.' Per lowered the sail a bit to slow their progress.

As they neared the rock, while Anna made a final effort to bail as much seawater as possible out of the cabin, Per studied the only piece of land in the otherwise vacant sea ahead. It sloped downward from left to right, and he decided the obvious course was to sail toward the low side. A rocky crag had appeared a few hundred metres offshore on the low side of the rock, but he judged that he could sail between the rock and the crag looking for a safe place to land on the rock.

As they drew near the rock's right side, a cove appeared. Unable to see around the corner of the rock beyond the cove, Per committed to beaching Manaia in the cove. 'Annie, do you see that cove?'

'Yes.'

'I'm going to try to hook to port inside the cove to get out of the waves. Hopefully there's a bit of beach in there. Stay with the boat unless I tell you to abandon it.'

'I understand.'

'We may run aground on a reef on the way in. If we do, we may have to abandon the boat and swim to shore. But do not abandon ship unless I tell you to.'

'I understand.'

The high tide floated them over what reef there was. Pulling in the starboard side of his jury-rigged sail and pushing the improvised steering oar, Per steered into the cove, able to hook to port as he had hoped, away from the stronger waves. To his immense relief, there was a bit of a beach.

Tūpuna Rock

Manaia howled as her keel hit bottom and the forward keel bolts were ripped through her hull, splintering the area through which they had passed. Per worked the steering oar furiously to turn the boat sideways so the remaining waves hit the hull broadside, pushing it farther up onto the beach. There were horrible crunching sounds as boulders on the beach dug into the surface of Manaia's hull, and then there was little sound but the lapping of the waves diminished by the windward corner of the cove.

Per studied the shoreline as he lowered the sail. It was clearly near high tide.

Too emotionally spent to express his tremendous relief, he simply proposed, 'Let's just sit here with our weight on the beach side of the cockpit for a while.'

As seawater poured in through the holes opened in Manaia's hull, she laid over onto the side weighted by her crew. After composing himself for a few minutes, Per eased himself over the side, helped Anna down, and they waded together to the first dry land they had set foot on in six days.

3   Ashore

Per and Anna sat side by side on the rocky beach watching the tide ebb. As Manaia lost the support of displaced water, a sharp chunk of basalt on which she lay pierced her hull with a cruel, groaning sound. Now parted from its rudder for many days, Manaia's useless tiller hung downward as a lifeless stick.

Ever since Anna and Per were very young, their grandfather had taken them fishing on Manaia during their holiday visits to New Zealand – 'Aotearoa', as their mother's parents always called it in Māori, 'the land of the long white cloud'.

Anna enjoyed the fishing. As her interest in mathematics grew, she amused herself by searching for correlations between her fishing success and any potentially relevant variable. Tables of rising and setting times for the sun and moon were packed in Anna's dry bag to complement the tide tables kept aboard Manaia.

From his earliest days, Per was more interested in the boat. When only seven or eight, still too small to see over the cabin to where the boat was headed, he begged his Māori grandfather, Koro, to let him steer. Noting the limits of Per's visibility, Koro made a game of it, trimming the sails to what seemed best for the desired heading and letting little Per use the tiller to find a heading that would fill the sails and get them underway. Unable to see forward over the cabin, Per looked aft, searching for signs in the water astern. At first, even with Koro's patient coaching, the sails would luff and flog or stall as the boat drifted aimlessly. In time Per saw a first sign. Leaning out over the transom, with his grandfather's hand hovering near the grab loop on Per's life jacket, the boy saw a weak eddy form at the stern. Soon, he watched the eddy begin to trail behind the sloop and then saw the first trace of a wake.

Per developed the feel of it and was thrilled as the boat surged forward in response to the helm. Once Per had the physical strength to haul in the sheet controlling the mainsail, Koro upped the game by trimming only the headsail and let Per manage Manaia's mainsail and tiller to best effect. When he was finally big enough to see over the cabin roof and reach the winches, still only eleven or twelve, Per had full command of the boat; Koro had nothing more to teach him.

Per had once seen a picture of King Olav V out sailing in the Oslo Fjord with his hand on the tiller of a sloop not unlike Manaia. Although sailing on the opposite side of the globe from the Oslo Fjord, Per began to see himself in that picture.

Koro said Per had a gift for visualizing air flowing around the sails and water flowing around the hull, keel and rudder. It was true that he had a gift; but he also had a passion. Per loved Manaia as much as anyone can love an inanimate object. He loved how Manaia's rounded hull sashayed over waves when heeled over, and how she punched through wind waves when sailing close to the wind. Most of all he loved how Manaia's tiller pressed or pulled his hand to tell him where she, herself, wanted to go. He loved the whole of her, but his first love was her now lifeless tiller.

As they sat on the beach, Anna watched her younger brother out of the corner of her eye, not wanting him to feel her watching. He seemed unmoved by the scene. Without a conscious thought, they had both moved into a survival mode. Although now on dry land, their long-term survival was not at all assured, and their brains, both conscious and subconscious, were focused on the next action necessary to keep them alive.

Per surveyed the beach, while Anna examined the cliff face behind them. The cove was about one hundred and fifty metres across, wide enough to embrace a football pitch, but not much more. The beach, at the current tide was no more than five metres wide and backed by a tall cliff. To the left, looking seaward, the beach vanished as the cliff reached out into the sea, blocking any exploration. To the right, the beach rounded a point. 'We've got to find a way off this beach,' Per warned. 'If the winds pick up again and blow directly into the cove, this beach is going to get hammered in the next high tide.'

'We've got a few hours before the next high tide. Let's take a look around that point,' Anna proposed, nodding toward the right.

Immediately around the point a narrow gully led inland and further beyond the point the beach narrowed to nothing as the cliff dropped straight into the sea. They returned to explore the narrow gully. A tight, barely passable, length of a few metres led to a flat spot two metres wide and five metres long, beyond that flat spot the gully narrowed to the point that it became impassable.

'What do you think? Is this a camp site?' Anna asked, glancing around the flat spot.

'It sure beats the exposed beach.'

They made their way back to Manaia, by then high and dry. Per climbed aboard, ignoring her hull's groans as his shifting weight pressed her against the rocks. He began to unload everything he thought they might need, cushions and sails to use as a tent or bedding, Anna's dry bag and fishing gear, Koro's tool boxes, first-aid kit, propane grill, the remains of the last fish Anna had caught and, perhaps most important, the twenty-litre water jugs Koro kept in the bilge.

'Why does Grandpa Koro keep so much water onboard?' Anna wondered aloud.

'He often sails her alone. Instead of having crew to move their weight to the windward side of the boat to counteract the wind's force on the sails, he keeps a couple of crewmember's worth of ballast down in the bilge in the form of these jerry cans full of water.'

'How many are there?'

'Eight. Eight twenty-litre cans. One hundred and sixty kilos worth.'

'We're going to be really glad we have those.

'And why are his tools onboard?'

'Koro and some mates are building a shed at the marina. He's storing his tools onboard until the shed is finished.'

'I suppose progress on the shed has stopped.'

'Yeah, but not because they're missing Koro's tools.'

That conclusion led directly to the thought of themselves being missing, and their family and friends processing their disappearance. That thought was a bottomless emotional hole. Although both could not avoid thinking about it from time to time, they did not voice the thought to each another.

As Per unloaded, Anna began to carry things toward their campsite. It was a peculiar landscape she traversed. The beach was a jumble of basalt blocks that demanded careful foot placement as she carried her loads. The beach itself was nearly black, while the ten-metre-tall cliff face was milky white pumice. Some ferns and scrubby brush had taken root here and there in the pumice layer, otherwise the scene was barren.

Before carrying away the last load of the day, Per freed Manaia's anchor from her bow roller and set it on the rocky beach at a length of chain. The beach didn't promise good holding, so he piled rocks on top of the anchor and hoped for the best.

They piled their supplies at the entrance to the narrow gully and then muscled them through to the flat spot together. The settee cushions were laid out in the flattest areas. Anna chose the soft nylon spinnaker to wrap up in while Per took a Dacron headsail. It was only mid-afternoon, but neither had had a real night's sleep in six days, and they were both soon fast asleep.

~ ~ ~

The rising sun of the following morning shone through the entrance of the narrow gully. Having slept through the remainder of the previous day and the whole night, Anna and Per both woke with the sun.

Per stood, took a sip of water from one of the water jugs, and looked out through the narrow passage to the sea. 'The tide is out.'

Anna retrieved the tide tables from her dry bag and, turning a page, offered, 'Oh, I forgot to wish you a happy New Year yesterday.'

Per faked a smile but said nothing.

Anna compared a tide table to her sun position data, 'Back in Whangārei low tide was an hour and a half before sunrise. I've got to start with the assumption that the sun and tides aren't so different here, wherever we are.'

'So best guess is that the tide is coming in.'

'We can watch the tides here and learn how they compare to those in my tables for Whangārei, but for now, the best guess is that the tide is coming in.'

'When's the next low tide?'

'18:17 – in Whangārei.'

'And sunset?'

'20:41 – in Whangārei.'

'So, if the times on this rock aren't too terribly different from Whangārei, we can explore this rock for most of the day and have another low tide to get back here before sunset.'

'Exactly. Let's go look for anything that might be of help, water, or some clue as to where in the heck we are.

'Before we go, I should change the bandage on your jaw. There's some fresh blood on it.' Anna opened the first-aid kit and carefully inspected the contents she had only scanned hurriedly in the dim light of Manaia's cabin while being tossed about by the storm. 'We're down to the last of the large sterile bandages.'

Per studied the kit. 'It's a good kit for cruising along the coast for a few days, plenty of stuff to keep you patched up until you return home.' He considered joking that it must now be time to return home, but realized there was no humour in it.

Anna peeled off the soiled bandage and examined the wound. The skin adjacent the five-centimetre-long gash was red. 'Uff da, looks like some saltwater soaked into the bandage. Does it hurt much?'

'Yeah. I was trying to ignore it; but, yeah, it hurts. I guess that means I've got less on my mind than I did while we were still afloat. Does it look infected?'

'I don't think so, not yet anyway, but I'm worried about that. I ought to clean it as well as I can.'

Per invited her to 'scrub away' and remained stoic as Anna washed the wound with disinfectant from the small bottle included in the kit and applied antibiotic from its little 30-gram tube. 'Looks like the cord cut most of the way down to our jawbone. I don't know what to do to avoid a scar.'

'Let's worry about infection. Having a scar doesn't seem worth worrying about now.'

As Anna applied the last of the large sterile bandages, she explained, 'Next time I can use two of the small bandages.'

'How many of those are there?'

'Just the two. There are a few little bandages we can use after that, and some gauze and tape, but that's all.'

'I saw some paracetamol in there. Is that the only pain reliever we've got?'

'Yeah.'

'How many are there?'

Anna counted the individually wrapped tablets, 'Twenty-four.'

'If I promise to be careful to not get hurt again, can I have two?'

Per washed the tablets down with some of Manaia's ballast water. 'The tide is coming in. A hike will take my mind off my throbbing jaw. Let's go see if we can find some help.'

Setting out to explore their 'rock', Anna and Per passed by Manaia, lying on her side, gouged and pierced by sharp chunks of basalt. As Anna watched her footing on the rocky beach, Per gave Manaia a long look but did not say a word.

Around the point at the far end of the cove, a rocky beach stretched off for nearly a kilometre. A few large driftwood logs, many smaller pieces of driftwood, some plastic beach trash and the wreckage of a floating dock had washed ashore along the beach. The washed-up dock bore Japanese markings. Per and Anna assumed it had been torn adrift by a tsunami or some other oceanic savagery, and that it gave no clue to where they were.

Several gullies led inland from the beach. One gully, near the far end of the beach, was particularly passable, and they chose to explore inland. After less than a hundred metres of uphill climb through the gully, they reached a plateau and most of their rock came into view. It was a bleak place, treeless, with nothing more than scrub brush, ferns and sedge grass taking root in the pumice. The sun, peeking through the clouds, confirmed their general sense of direction; they were on the eastern edge of the rock.

As they had observed from the sea, the plateau sloped upward from east to west, and they set off for the high point on the far side. They hiked close to the northern edge of the plateau looking for signs of life on the side of the rock they had not seen as they had sailed up from the south. Per had secretly hoped an automated lighthouse would suddenly come into view as they reached the summit. 'No buildings,' Per began, 'no foot trails, not even any game trails.'

'It's just us and the birds,' Anna noted the plentiful birds that had skittered through the brush and occasionally taken flight as they walked along. 'It doesn't seem like anyone has been here for a long time, if ever.'

'Polynesian voyagers must have known this place,' Per declared. 'They explored every rock in this ocean.'

The summit proved to be the crater rim of a dormant volcano. Water had not collected in the crater; it was, like the rest of the rock, bone dry.

It was a pleasantly warm day with modest eight-knot winds coming from the south. They stood on the rim and took in the view.

Tūpuna Rock

Thinking of the ancient voyagers, they both searched the horizon. Under the broken cloud layer, the visibility over the surface of the sea was very good. The twin rocks they had passed two days earlier could be seen in the distance to the south. No other land was visible in any direction. The pinnacle they had passed earlier, like every habitable patch of planet Earth, was beyond the horizon.

The vast expanse of sea had a particular impact on Per. Under other circumstances he would have found it perfectly beautiful. He had sailed seemingly countless kilometres in the Canadian Salish Sea and along the coast of New Zealand, but he had never before sailed out of sight of land, and he had always known a compass bearing to sail to safety if fog rolled in to obscure his view. Now he had no compass and knew no bearing to safety.

Anna took in the horizon as well, finally turning back to Per. She was accustomed to seeing him scan the horizon, his head turning slowly as his gaze panned across the scene. But his head remained motionless, he was frozen with his eyes fixed on some faraway point, beyond the horizon. 'Per?' she asked quietly.

He did not respond.

Once again she asked, 'Per?' Again getting no response.

Finally, she stepped in front of him, facing him, looking into his eyes as they remained focused, through her, on the point beyond the horizon. She raised her arms, resting her hands on his shoulders and gave them a gentle squeeze. 'Are you okay, Per?'

He blinked.

Once again, she asked, 'Are you okay, Per?'

Per blinked again and, still staring ahead, his mouth moved, finally quietly forming the words, 'I'm scared.'

'It's okay. I've been scared for a week.'

'It's worse.' He blinked again, and his focus finally came to Anna's face as he spoke slowly. 'There was always something to do that might help. Secure something, cut something free. Sheet a sail in, let one run out. Maybe it didn't actually help in the end, but at the time it was something that seemed like it might possibly help. There was something to try.

'But now there's nothing,' he continued. 'We have no idea where we are, no idea which way to go if we could go. We have no way to leave this rock, anyway. Manaia will never float again. Noone is looking for us.' He paused before concluding, 'I'm scared.'

'I'm scared too, Per,' Anna assured, her hands still resting on his shoulders. 'We'll think of something.

'There was a branch of the gully we came up from the beach that looked like a better camping spot than the one we've got,' Anna offered. 'Let's go back and check it out.'

'Yeah, let's check that out,' Per agreed, regaining some life as Anna took him by the hand and led him off, back across the plateau.

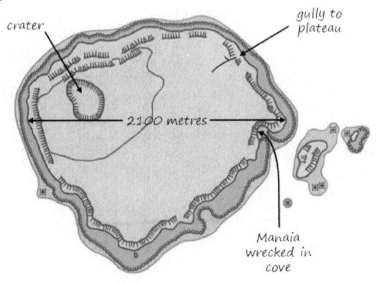

The gully branch proved to be an acceptable camping spot. It was shorter, draining a smaller area of the plateau than the main gully, so less likely to flood in a downpour. At the same time, it offered easy access to both the beach and the plateau through the main gully. They adopted it as their new home.

As they sat in the gully branch waiting for the tide to ebb, Per studied the walls and floor of the gully, 'They camped here.'

'Who.'

'The voyagers who came to this rock.'

'How do you know? Do you see a carving or something?'

'No.' Per could see that Anna was worrying about him. 'I'm not flaking out and seeing ethereal beings with moko tattoos on their faces floating around doing a haka,' he assured. 'But it is what they would have done. Maybe they landed here to make repairs, look for water, or just explore. Maybe all of those things at different times. But their spirits of exploring, of navigating and voyaging are here. I can feel it.'

'Hold on to their spirits, Per.'

~ ~ ~

By mid-afternoon the tide had ebbed enough to give them a narrow, walkable path on the beach around the point to the cove. They began the process of moving everything they had positioned in their first campsite to the gully branch. It was slow going on the rocky beach.

As they rounded the point on one trip, Per handed Anna her tackle box and let her choose a rod, 'I'll take the rest. You should get to work.'

Anna scanned the topography of the beach and the crag two hundred metres off the point. Thinking that the remains of the big haku she caught two days earlier would soon become unfit to eat, she chose a light casting rod and a small lure, not wanting to catch larger fish than she and Per could eat in a day. She projected that a reef extended at least part way from the point to the rocky crag offshore, waded carefully to knee-deep water off the point and cast the lure out into the channel between the point on which she stood and the crag. The cool water flowing around her legs was comforting, and she was in her element, fishing.

When Anna returned to their new campsite, Per was making a primitive tent over the gully branch, stretching the genoa, Manaia's largest headsail, between shrubs high on opposite walls. 'What are those?' he asked, studying the four trout-size fish she carried. 'Some kind of catfish?'

'Koro calls them "āhuruhuru". He says they're tasty.'

'I'll make a fire.'

With no idea how many meals they would have to cook on the rock, Per decided against grilling the fish on the propane grill from Manaia. Mentally comparing one of the big propane cannisters to the size of a little finger-size pocket lighter, Per was reassured to think that, if he just used the grill as a fire starter, lighting small bits of driftwood kindling and then promptly shutting the grill off, one cannister could last for months, if not years. They had two full cannisters, and a third nearly full.

At the mouth of the main gully, near the beach and a supply of driftwood, Per built a pile of rocks in the shape of a horseshoe with its sides close enough together to support the ends of the grill from Manaia's oven. He lit a small pile of twigs with the propane grill and then added larger pieces of driftwood to build a cooking fire.

As Per built the cooking fire, he began a separate pile of driftwood sticks on the beach near the entrance to the main gully. He pulled sedge grass and brush from the hillside and left it to dry under the sticks as the start of a signal fire to be lit as soon as a ship appeared.

The āhuruhuru were, indeed, tasty. Three of the fish were enough to satisfy the siblings and they wrapped the fourth in a piece of sailcloth cut from the top of the mainsail.

Per leaned back against the side of the gully, 'I haven't heard an airplane, seen a ship or heard anything on the radio. Why isn't anyone searching for us?'

'I was thinking about that as I was wading in the current, fishing. You know how the current runs clockwise in the North

Pacific, up the coast of Asia and then down along the west coast of North America.'

'Yeah.'

'It would be the opposite here in the southern hemisphere, counterclockwise, up the coast of South America and down past New Zealand and Australia. They must think we've been carried by the southbound current, not blown by the wind.'

'Wreckage in the current.'

'Yeah.'

'So, if we hadn't kept Manaia afloat, we would have been found.'

'If you hadn't kept Manaia afloat, they might have found our drowned bodies drifting toward Antarctica.'

'So being on this rock isn't the worst thing.'

'Definitely not the worst thing. We've got shelter,' she looked up toward the sail overhead. 'Will rain run off this?'

'The sailcloth? It's tightly woven. Yeah, rain will run off.'

'How big is it?'

'About twenty square metres.'

'So, if we catch the runoff, a centimetre of rain can give us two hundred litres of water. We've got shelter. We can get food and water…'

Per smiled at his sister's instant calculation of litres per square metre. 'Fish is great. But we may get tired of nothing but fish.'

'What would the ancestors, the tūpuna, the ancient voyagers who must have been here, do?'

'There's a lot of fern growing on this rock. Wasn't baked fern root a thing?'

'Okay, we're not drifting toward Antarctica. We've got shelter, ways to get food and water. What would the tūpuna do next?'

'They'd plot a course to Aotearoa and sail home.'

'Sail what? Manaia is beyond repair, certainly not with what we've got to work with on this rock.'

'If it's not too far, maybe we could rig your Viking sail on that piece of floating dock, wait for the right wind and sail to Aotearoa.'

Per was not at all keen on the idea of sailing a piece of broken dock. 'If it's not too far,' he echoed. 'How would the tūpuna figure out if it was too far? Hell, how would they figure out which way to go?'

Anna moved to the edge of the sail canopy and looked up at the cloudy sky, 'They'd look at the stars and figure out where they were.'

'I don't know how they did that without a sextant, chronometer and nautical almanac, but you're right, they'd look at the stars and figure out where they were.'

Like Per, Anna was beginning to feel the stirring of a spiritual connection to their mother's ancestors, the tūpuna, the voyagers who had visited the rock they found themselves on. As if on cue, wind rustled the brush growing around their shelter. 'We can figure it out Per. The tūpuna will teach us. Their spirits will show us something in the stars.

4 Latitude

'How did the Vikings determine their latitude at sea?' Anna wondered aloud as they sat in their gully-branch campsite.

'I don't think they did.'

'Huh?'

'From what I've read, they sailed up the Norwegian coast to the latitude of their destination – Trondheim works for sailing to Iceland. Ålesund is good for going to southern Greenland. While they provisioned in those harbour towns, they'd note the height above the horizon of the noonday sun and, at night, Polaris, the north star. Maybe they also watched a zenith star, a star that passes directly over Trondheim or Ålesund. Then they'd set sail westward, keeping the height above the horizon of the sun at noon and Polaris by night just as they saw them back on the Norwegian coast, and watching for their zenith star to pass overhead. If the sun or stars weren't as they'd seen them back in Trondheim or Ålesund, they shifted course until they were as they remembered, and they'd know they were be on the latitude that took them to Iceland or Greenland.

'I don't mean to make it sound easy. It was really sporty for sure, but I understand how they did it. They didn't have to determine their latitude at sea, they just sailed on the latitude of Trondheim or Ålesund all the way to their destination, watching the heights of stars above the horizon.

'We don't have a coast to sail down, or up, to a place we know; there's nothing but water out there. And we wouldn't know whether to turn westward or eastward when we got to that nowhere in the middle of the ocean.'

Anna took it all in and thought for a long time. Suddenly she couldn't control her excitement, 'Wait! Remember how Karani always took us down to the beach before sunrise?'

'Of course,' Per replied with a chuckle.

'She said the ancient Polynesian navigators knew that, if you made landfall on Aotearoa where the lowest star of Māhutonga, the Southern Cross, or the star Marere-o-tonga were one-and-a-half thumbs above the horizon at their lowest points, the Ngāti Te Taki-o-Autahi, her tribe, her iwi, our iwi, were home.'

'Our iwi?'

'Āe, Pita,' she said 'yes' in Māori and called him by his Māori name, 'pee-tah'.

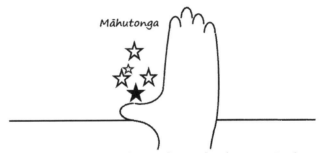

Māhutonga is one-and-a-half thumbs above the horizon, where Ngātiwai are home.

'Karani was teaching us to recognize in the sky over Whangārei just what those Viking navigators saw in the sky over Trondheim or Ålesund, the latitude of their destination.'

'But we still wouldn't know whether to turn eastward or westward. This is a much bigger navigational problem than our Viking ancestors had to solve. Latitude – north-south – is relatively easy. Longitude – east-west – is really hard.'

Anna was undeterred, 'I think our Viking ancestors can help along the way, but we've got a tougher problem in a different ocean. If we don't think like Polynesians, like Ngāti Te Taki-o-

Autahi, maybe one day someone will figure out what happened to us when they find our bones on this rock. I'd rather get back to Aotearoa and tell how we did it.

If we tap into our Ngāti Te Taki-o-Autahi heritage and learn from the spirits of our ancestors, the tūpuna, I believe we'll figure out where we are and then find our way home to Aotearoa.'

'Āe, Ani,' Pita replied, pronouncing her nickname, 'ah-nee', as in Māori, signalling his agreement. 'So, the tūpuna knew how to use Māhutonga to determine latitude – how far north or south they were. How does that work?'

'Māhutonga circles around the southern celestial pole. The farther north you are, the less space there is between Māhutonga and the horizon. At the extreme, back in Canada or Norway, Māhutonga is so far below the horizon you can never see it. A little less extreme, in Hawai'i you can only see it when it's at its highest. On the other hand, on the South Island I think you could fit your whole hand between the horizon and Māhutonga at its lowest point.

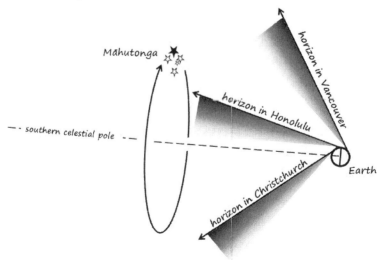

Ani checked the chart they had for the approach to Whangārei harbour and their fishing grounds, 'Karani's beach is at about 36 degrees south latitude.' Then she sketched in the notebook kept in her drybag.

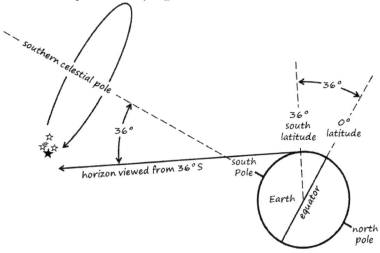

'When we see Māhutonga at its lowest point from Karani's beach, we're actually looking over the south pole. That's why, at its lowest point, Māhutonga is upside down, relative to how it's normally pictured, how it is on the Aotearoa New Zealand flag.'

'Karani taught us to measure north-south latitude with our thumbs,' Pita noted. 'What's a thumb, exactly?'

'Karani uses her thumb to measure angles, like we use degrees. The tūpuna may have done the same,' She sketched some more. 'I think I'll stick to math and not consider a future as an artist; but you'll get the idea.'

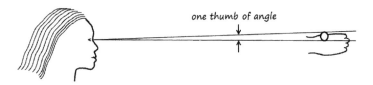

'That's actually not bad. You captured your large Polynesian brain case and even your Viking blonde hair. But how much is a thumb of angle?'

Tūpuna Rock

'Measure the distance from my eye to my thumb.'

Pita used a folding rule from his grandfather's tool box to measure. '51 centimetres.'

'So, if I could swing my thumb around my eye, the circumference of the circle would be 51 centimetres times 2, to get the diameter of the circle, and that times pi.' After a little mental math, she concluded, 'The circumference would be 320 centimetres.'

Pita chuckled at his sister's mental math.

'How wide is my thumb?' Ani asked, holding out her thumb.

'It's 2 centimetres wide.'

'Okay, so a full circle is 320 divided by 2, 160 thumbs. One thumb of angle is one 160th of a circle.

'The width of your thumb and the length of your arm, and certainly those of little Karani will be different from mine. But the proportions, like Karani has both shorter arms and skinnier thumbs, will be similar. That's all that matters, a thumb of angle will be similar for all of us, about one 160th of a circle.'

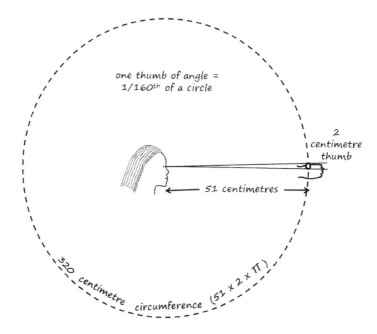

'Do you remember the circumference of the Earth?'

'Forty thousand kilometres, isn't it?'

'Yeah, the whole basis of the metric system is the circumference of the Earth around the poles – 40,000 kilometres. So, a thumb of latitude is 40,000 divided by 160. Two hundred and fifty kilometres.'

'So if we find that Māhutonga is only a half of a thumb above the horizon here, one thumb less than in Whangārei, we're 250 kilometres north of Whangārei.'

'Yup.'

Pita observed, 'That's about a day's sail in a decent boat with average winds.'

'Really? Maybe that's how the tūpuna thought about a thumb of latitude, as a day's sail.'

By day, as they waited for a clear night to view Māhutonga, they looked for food sources on the rock. Gathering eggs from the plentiful birds was easy. After finding a nest on the plateau, Pita would wave a piece of the orange storm jib to shoo angry birds away while Ani plucked an egg from the nest. The eggs were only slightly smaller than a chicken's egg, and they had a small pot half full in less than an hour.

After returning to the gully, they put a frying pan on the cooking fire's grate and began breaking eggs over it. Cracking the first egg revealed a nearly fully developed chick with a bulging skull and scrawny wings all covered by tiny wet feathers. Pita recoiled at the sight, 'Ugh! Gross!'

All but one of the eggs contained a chick in some stage of development, a few had a bit of edible yolk sac remaining alongside a developing chick. 'I guess they keep the roosters away from the hens at egg farms, so the eggs aren't fertilized,' Ani surmised. 'The birds may only come here to breed; they won't be very interested in a vow of celibacy. So much for eggs.'

They could have fried the small puddle of edible egg in the pan, but the pile of feather-covered bony wings and skulls set to the side had killed their appetite. Ani carried the carnage to the beach and dumped the pile into the water, hoping it would nourish some sea life.

Turning their thoughts to ferns, Pita warned, 'I don't remember it exactly from the Outdoor School back in Vancouver, but there was a lot of talk about ferns being poisonous.'

'Some must be okay here in Aotearoa. It was a Māori staple once. Karani once dug one up for me and told me how they prepared it.'

'There are different kinds of ferns here. Could you recognize the right one?'

'Yeah, I think so. I'd recognize the roots for sure, once I saw them.'

They took two pieces of driftwood to use as digging sticks and returned to the plateau. Digging at the base of the big ferns most common on the rock, Ani found their thin roots looked nothing like those her grandmother had shown her. Finally, after digging at a few other ferns, she found roots exactly like those Karani had shown her and began harvesting.

Meanwhile, Pita had taken some monofilament fishing line and begun setting up snares to catch birds. Taking a two-metre length of the line, he tied one end to a shrub near a nest and tied a small loop in the opposite end. Twirling the small loop between his thumb and index finger, he pulled enough of the line through the small loop to form a large loop that would close tightly around anything caught in it. Finally, he hung the large loop between shrubs near the entrance to a nest. He repeated the process at several more nests before joining Ani to help harvesting fern roots.

In less than an hour, all hell broke loose when a petrel was caught in one of the snares. Pita grabbed his digging stick and ran toward the leaping and flapping bird. Before Pita could reach his quarry and subdue it, the bird lifted off, trailing a branch of the shrub at the length of the invisible snare. 'Crap!' was all he could say as he watched the bird and branch fly away.

'What happened?'

'A bird got caught in a snare and ripped off the branch it was tied to.'

'Where's the snare?' Ani demanded.

'Out there,' he said pointing out to sea, 'still tied to the bird and the branch.'

'Damn it, Pita. It's gonna end up in the ocean somewhere.'

Loose fishing line is a scourge, entangling and choking all forms of marine life as it floats around the ocean for decades. Dangling in the water from a floating branch, Pita's snare could continue killing indiscriminately for years.

Ani was religious about making sure none of her line got away. Beginning to steam, she demanded further, 'How many of those snares do you have out there?' as she swept her arm across the plateau.

'Six… …five now.'

'Gather them all up right now! We'll eat fish for protein.'

After nearly capturing the petrel, Pita felt he was on the verge of success, but he could see in Ani's eyes that it was not something she was willing to discuss. He reluctantly retrieved the remaining snares, and the siblings retired to their campsite with a bundle of fern roots which Ani spread out in the sun to dry.

~ ~ ~

By late afternoon the sky had fully cleared, and Ani shuffled through her tables. 'We can see to the south from the point, and we've got a low tide at sunset. We can go to the point tonight and measure the height of Māhutonga above the horizon.'

They sat together on the beach at the point, patiently waiting for stars to appear as the daylight faded. It was a few days before the full moon and the moon was already high in the sky as the sun set. Blocking the bright moon with their hands, they saw the first bright stars appear in the east. Both siblings recognized Orion's belt, and Pita, the sailor, knew Sirius, the brightest star in the night sky. At last, the stars of Māhutonga began to appear in the southern sky. Their hearts sank when they realized that Māhutonga was past its low point and already rising in the southeast when it became visible. 'Maybe tomorrow?'

Ani thought for a moment. 'It was during our July – August school vacations that Karani showed us the height of the Southern Cross – Māhutonga. She'd wake us before dawn and take us down to the beach. She always did that on the first clear night as soon as we got to Aotearoa in July. She said all the stars reach their lowest point a few minutes earlier each night. If we had waited until August, we'd have had to get up in the middle of the night to see if we were in our Ngāti Te Taki-o-Autahi homeland.'

'So now it's just going get worse each day. Māhutonga's low point is just going to get earlier and earlier before sunset each day, when it's daylight and we can't see stars.'

'Until July when all those minutes earlier add up so we can see it just before dawn.'

'That's six months from now.'

'What now?'

Ani paused, realizing that finding their latitude was not going to be as straightforward as she had hoped. 'We sight something to use as a comparison.'

Ani held out her hand with her thumb and pinkie stretched far apart. 'Do you see that bright star about a handspan and a half to the right of Māhutonga?'

Pita held out his hand, mimicking Ani. 'Got it.'

'Find a piece of your hand that fits between that star and the horizon.'

They both turned their hands this way and that, while held at arm's length. 'The width of my open palm near the base of the thumb is close.'

'Mine, too. It may still be getting closer for a while. Keep checking and note how close it gets.'

They both raised an open hand every ten or fifteen seconds until they agreed the star was rising again. 'At its closest, it was just a few millimetres wider than my palm.'

'Mine too. There's our data point. Do you have any idea what that star is called?'

'Not a clue. How do you say "happy new year" in Māori? It's "ngā mihi o te tau hou" isn't it?'

'Āe, let's call it "Tau Hou", "New Year".'

'So, at its lowest point, the star Tau Hou is one palm plus a few millimetres above the horizon.'

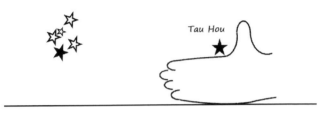

'The stars move in an arc. How will we compare the height of Tau Hou to Māhutonga?'

'It'll come to us.'

~ ~ ~

The following day they were carrying what remained aboard Manaia and not bolted in place to their campsite. Ani was carrying the oars to the lost dinghy. Pita glanced at the oars and, noting that they were straight, an idea came to him. One could be an artificial horizon set to mimic the horizon an any angle that let them observe the passage of both Tau Hou and Māhutonga. The mathematician could figure out the best angles.

Ani loved the idea. To keep things relatively simple, they could set the oar artificial horizon so it appeared vertical while they viewed the stars as they swung to the left of the celestial pole. As long as both stars were measured to the same artificial horizon, the angles didn't have to be precise. Since Whangārei was about 36° south and they were confident they were north of Whangārei, somewhat more than 30° would be a good place to start.

Māhutonga's farthest swing to the east would be a bit more than 30° east of south. They should set their artificial horizon so it was 30° or so east of south from the observer.

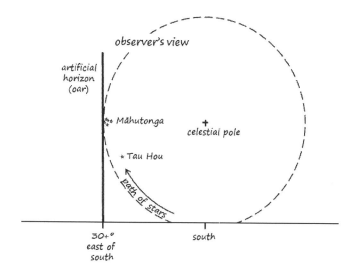

Sighting the stars as they were level with the celestial pole, they would be looking upward, at the same angle as the surface of the Earth to the celestial pole – tipping the artificial horizon oar back by 30° would work just fine.

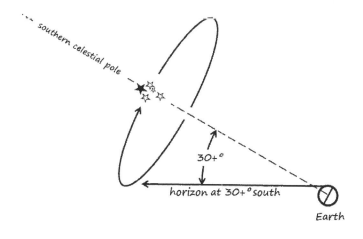

'That's a lot of thirty-degree angles to figure out,' Pita noted as they made their way down to the beach to gather driftwood to build their observatory.

'Thirty degrees is the easiest of them all.'

'Lay it on me.'

Ani knelt down and drew a triangle with three equal sides in a patch of sand. 'Remember what the three angles will be if the sides are equal?'

'Three equal sides mean three equal angles. The three angles must add up to 180 degrees, so each one is 60 degrees.'

Ani drew a line from one angle to the middle of the opposite side. 'And those angles?' she asked pointing to the bisected 60-degree angle.

'They're both 30 degrees.'

'And if each side of my original triangle is two metres long, how long would the short sides of the new triangles be?'

 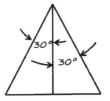 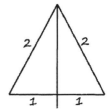

'Jeez, I thought half of sixty degrees was going to be the easiest question. Half of two metres is one metre. So, if an oar is two metres long, and we hang a weighted string from the top, when the bottom of the string is one metre away from the bottom of the oar, the oar is at a 30-degree angle.'

'Exactly. I mean both that you understand, and the angle will be as exact as anything we're able to do on this rock, far more precise than it needs to be.'

'Is this all what the tūpuna would have done?'

'They knew the stars better than we ever will. Kupe, the great navigator, is looking down from Father Sky at those triangles,'

Tūpuna Rock

Ani pointed to her drawing. 'He is smiling. Maybe he made the same drawing in a patch of sand to teach our tūpuna centuries ago. It's a sure thing our tūpuna didn't measure angles in degrees, maybe they used thumbs at arm's length, or something else. But stars and Mother Earth, and the angles between them... ...they knew them intuitively from observation and experience and from the teachings of their own tūpuna. We've got some catching up to do. I'm sure they thought up something at least as good as the observatory we're building.'

'Ani, do you think the spirits of the tūpuna that live on this rock gave me the idea for using the oar as an artificial horizon?'

'Yes,' she answered without hesitation. 'I don't say that to take anything away from you. Maybe the tūpuna spoke to you through your genes, maybe their spirits spoke to you. What matters is that you were listening, and you heard them.'

On the plateau, with the oars, driftwood sticks from the beach and line from Manaia, they built the 'observatory' they had dreamt up. They set up the second oar as a sight bar, lashing two driftwood tripods together to support it parallel to the ground and perpendicular to their line of sight 30° east of their best estimate of south. To make their observations easy to convert into thumbs of angle, they placed the sight bar 2.55 metres back from the vertical oar – five times the distance from Ani's eye to her thumb held at arm's length. Around the shaft of the second oar, they tied two bits of string that they could slide along the shaft to mark the positions of Māhutonga and Tau Hou.

Father Sky's view

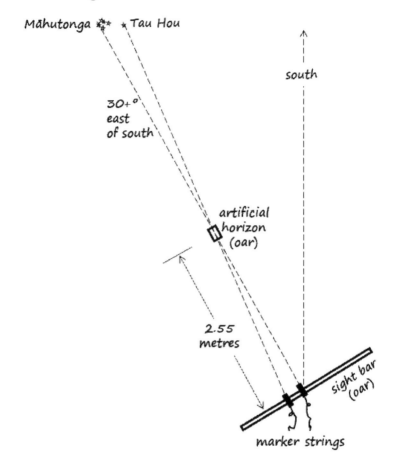

It was pleasantly warm, but the wind had picked up, so with the observatory set up, they retrieved sails to wrap up in while they waited for the stars to orbit the celestial south pole. The near-full moon was high in the sky as Māhutonga rose in the southeast, giving Ani and Pita enough light to work by without compromising their view of the rising stars.

Ani patiently moved the marker string for Māhutonga to the right as the constellation swept around the celestial pole. It was the middle of the night before Māhutonga reached its easternmost point. Pita, the night owl, took over to track Tau Hou's eastward progress, inching its marker string rightward as it went. The first light of pre-dawn twilight was already apparent on the horizon when Tau Hou ceased its eastward progress. Ani, not known as a night person, had remained awake through the whole process, excited by the unfolding data. Despite the promise that daylight would soon enable a safe return path to their campsite, they wrapped up in sails and slept beside their makeshift observatory.

~ ~ ~

'What do you think, sis?' Pita asked as soon as Ani began to stir.

'Someone once said, "The data will set you free." I feel liberated. Do you have Koro's folding rule and tape measure?'

'Right here.'

Ani crawled out from the folds of sailcloth and proceeded to measure the distance between the two marker strings and confirm the distance between the two oars, the artificial horizon and the sight bar. 'Let's go get some breakfast.'

'All that work, building and staying up all night and you only want two measurements?' Pita challenged with a note of playfulness.

'It's all we need.'

They ate some fern root, for which they were developing a tolerance, if not a taste, and the remaining āhuruhuru from the night before.

'We've got to learn to smoke fish, to preserve it. This seems fine, but I wouldn't want it to go much longer.'

'Can we first figure out what we learned last night?'

Ani looked at her notes, 'We located the sighting bar 2.55 metres from the artificial horizon – five times the distance from my eye to my thumb. It was 525 millimetres between the marker strings. One fifth of that 525 is 105 millimetres. If we had only been looking at the artificial horizon from the distance of my arm's length, the marker strings would have been 105 millimetres apart.

'My thumb is 20 millimetres wide. So, Tau Hou at its lowest point is, 105 divided by 20 — five-and-a-quarter thumbs higher than Māhutonga.

'The night before last we saw Tau Hou one palm above the horizon. How wide is my palm?'

Pita measured 80 millimetres. 'Divided by 20 that's four thumbs above the horizon. And last night we learned that Māhutonga is five-and-a-quarter thumbs lower than Tau Hou.

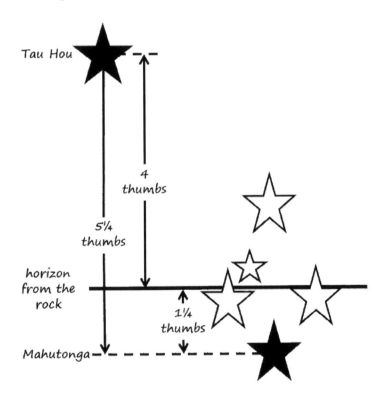

'We wouldn't have seen the lowest star of Māhutonga from the beach, even if it hadn't been daylight. That star was, 4 minus 5¼ — one-and-a-quarter thumbs below the horizon.

'And on Grandma Karani's beach Māhutonga is one-and-a-half thumbs above the horizon.

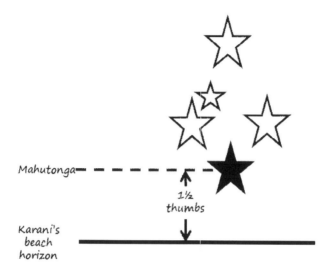

'So we're 1¼ plus 1½ — two-and-three-quarter thumbs north of Whangārei.'

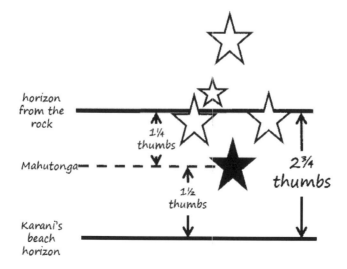

Pita was crestfallen. 'That's nearly seven hundred kilometres,' he observed, voice filled with disappointment. 'No wonder we don't hear anybody looking for us.'

'Yeah, and that's only the northward bit. We still don't have a clue how far east or west we are.'

'So, even if we knew which way to go, Aotearoa is three or more days' sail away, maybe a lot more, in a real boat.' Pita shook his head. 'We really can forget about trying to sail that piece of floating dock.'

5   Longitude

Having no artificial light, they quickly adjusted to the sun's day, going to sleep shortly after sunset and waking with the dawn. The tide was low at dawn on their first days on the rock, so they made morning trips around the point to Manaia to recover whatever they could. By the third day they had removed everything that was not bolted down and began unbolting everything they thought they might possibly need.

Removing the chain plates, heavy stainless-steel bars that had secured the mast shrouds to the gunwales, Ani was in the cabin loosening the nuts before Pita turned the bolts from outside. Pita had to pull his wrench so hard to break the bolts free from adhesive sealant that Manaia rocked subtly with the effort.

Out of the corner of her eye, Ani noticed a beam of light dancing across the vee-berth door. Sunlight coming in the open hatch was reflecting off the ship's clock. The clock had filled with water when Manaia was rolled, stopped and still showed the time of the event. As Ani and Pita were removing the chain plates, the sun's light was reflecting off the clock's lens onto the door. Ani was amused by the thought that they could put marks around the interior of the cabin to note the passage of time. 'Pita,' she suddenly shouted, 'I have an idea. Can the rest of the chain plates wait until tomorrow?'

'Sure,' Pita answered, poking his head into the cabin to find Ani removing the door from its hinges. 'Do you need help?'

'Don't think so. It's going to be a slow process, so you might just want to keep working here.'

'Okay. Need anything else?'

Ani sat back and thought quickly. 'I'll need the level, some long bolts or nails that I can drive into the ground… I think that's all. I've got plenty of fishing line in my tackle box back at camp.'

'Sure you don't need any help?'

'Thanks, but no, I don't think so.' She stuck her head out through the hatch looking up at the sun and then added, 'I'd explain, but that would take a while. I need to get started as soon as I can.'

Ani handed the door out through the hatch and climbed down over the side. Pita handed over the door and Koro's spirit level, along with a handful of long bolts and pins.

'I'll be up at the observatory,' she called back as she hustled toward the point with the door under one arm, clutching the level in one hand and the bits of hardware in the other.

She fetched a few things from camp, her phone, fishing line, her notebook and a pen. Saying a small prayer that the battery hadn't died yet, she turned on the phone and smiled when the time appeared on the screen. The smile turned to a frown when she read the time, 11:46. She needed readings well before noon on a sunny day, and after so many cloudy days she had no idea when the next sunny day would come. Looking up at the sun, she thought it couldn't be noon yet and realized her phone was showing daylight saving time. Father Sky knew it was actually about an hour earlier. Still, she had no time to lose.

Back at the makeshift observatory, she laid the door down under the sloping artificial horizon and held her finger over it. The shadow of her fingertip was fuzzy, making it impossible to accurately locate the end of the shadow. When she formed a circle with her thumb and index finger, its shadow was also fuzzy, but it was easy to accurately locate the centre of the fuzzy circle. She quickly broke off a fern and stripped it down to the stem. She wrapped the end of the stem around her finger and, after pulling it off her finger, tied it in a ring with fishing line. Finally, she tied the ring to the artificial horizon a metre and a

Tūpuna Rock

half above the ground, adjusting it so it cast a round shadow on the door.

She shifted the door so it ran east-west and the shadow of the ring was near the northwest corner. A few broken bits of brush under the edge were all that was needed to get the door level, and bolts and pins driven into the ground at the door's edges kept it from shifting. She quickly marked the centre of the ring's shadow on the door, turned on her phone and noted the time, 12:06. She left her phone on. This was a make-or-break experiment, worth exhausting every milliwatt-minute that remained in the phone's battery. She only hoped it would last until well past solar noon.

Every few minutes for nearly two hours she marked the centre of the ring's shadow and noted the time. It was a glorious day, pleasantly warm with a gentle breeze as she knelt over her addition to the observatory in her shorts and bikini top.

Working methodically in the minutes between readings, she tied a length of fishing line to a small rock, passed the end through the ring and gently lowered the rock to the ground, marking the spot directly below the ring where she drove a pin into the ground. A length of line was tied to that pin and a knot tied in the free end to mark the distance to the 12:06 mark. Finally, she swung the line around toward the opposite end of the door and marked an arc where the knot passed over the door's surface.

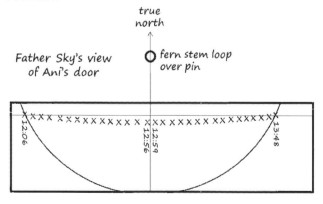

When the line of marks finally reached the arc, she noted the time, 13:48, and finally turned her phone off. That was it. Solar noon, when the sun was due north, was at 12:57, half way between 12:06 and 13:48. As a check, she drew a line from the 12:06 mark to the 13:48 mark and added another line, perpendicular to the first, at its centre. That line passed between the marks from 12:56 and 12:59, closer to the 12:56 mark. 12:57 was the best estimate of solar noon she would be able to make. With that single data point in mind, Ani hiked down to the camp.

Pita was on the beach gathering firewood and rocks to build a smoker as Ani studied the sun tables for Whangārei she carried in her dry bag. Solar noon didn't occur in Whangārei until 13:26 that day, 29 minutes later than on their rock.

~ ~ ~

As soon as Pita returned to the tent, he noticed Ani's shoulders and back, which were turning crimson from exposure to the noonday sun. Thinking immediately of single small tube of burn cream in the first-aid kit, he became furious at her carelessness. 'Damn it! What the crap were you thinking?'

'I'm really sorry. I got caught up in my data gathering. I guess the old "hurts me more than it hurts you" won't excuse much.'

'It's not about who hurts more. We're in this together, Ani. We need everything in that tiny first aid kit for real accidents, not to make up for careless sunbathing while doing some nerdly experiment,' Pita heaped on the accusations.

The 'nerdly experiment' accusation stung more than the sunburn. Ani sulked while Pita seethed.

A half hour of silence passed before Ani thought they could move on. 'Can I give you a test?'

Pita gave Ani a quizzical look, before asking cynically, 'Do I get to study? Do I need the calculator?'

'No studying beforehand, but you can use the calculator for some questions.' Ani intoned her voice to try to brighten things up.

'Okay,' curious, Pita loosened up a little.

'First, where does the sun rise?'

'In the east.'

'Good so if it rises earlier in one place than another, or if it's noon earlier in one place than another, where is that first place relative to the second place?'

'To the east.'

'Good. That's because the Earth is spinning under the sun. How fast is the Earth spinning?'

'One revolution per day.'

'In minutes?'

'Really? Are you going anywhere with this?' Pita, still on edge, demanded. 'Or are you just trying to distract me from your sunburn?'

'Stick with me. This is going somewhere.'

'Okay.' Pita reached for the calculator and punched in 24 hours times 60 minutes, 'One revolution every 1440 minutes.'

'Great, now how many thumbs of angle in a revolution?'

'160. See, I remember.'

'Okay, how many thumbs does the Earth rotate per minute?'

Pita punched 160 thumbs divided by 1440 minutes into the calculator, '0.1111111 .'

'You're doing really well,' Ani chuckled at her simple test. 'Just one more question. I've been up at the observatory determining solar noon at this rock. I have sun position data for Whangārei. Solar noon is 29 minutes earlier here than in Whangārei.'

Pita almost leapt to his feet in excitement, 'We're east of Whangārei, still in the Pacific!' Pita paused. 'Wait a minute,' 0.1111111 was still showing on the calculator, he hit, * 29 =, 'We're 3.22 thumbs east of Whangārei.

Mentally adding the eastward 3.22 thumbs to the northward 2.75 thumbs, Pita's mood turned sober. 'It's not an answer I'd hoped for. It's too far.'

Pita was silent for a full minute, contemplating a thousand or more kilometres of open ocean separating them from safety. 'But it is an answer.'

Eventually trying to brighten up the mood he asked, 'How did I do on your test?'

'I'd call it a solid "A+".'

'Can I have a prize?'

'We don't exactly have a cache of presents.'

'I know something you could do for me.' Pita unfolded the light blue spinnaker, took out his rigging knife and cut a metre-long triangle off the head of the sail. He folded the top half metre down, punched holes at each end of the fold with the marlinspike on his knife and tied a short piece of cord to each hole.'

Ani was puzzled by the production. 'What is it?'

'It's my version of a sun cape, kinda like Karani's hikurere. The cords are to tie it around your neck. Here's what I'd like as a prize. I want you to wear this any time you work out in the midday sun.'

A warm smile spread across Ani's face. 'Thanks bro, I will. I promise. But this is a prize for me. What about for you?'

'Knowing where we are is the biggest prize this skipper could get.'

'By the way,' Ani explained, 'when I thought I was wishing you a happy New Year a day late, it actually was New Year's Day. We're across the date line, back on the Canada side.'

A sudden gust of wind blew one edge of the genoa covering their encampment loose. Some of the shrubs the sail had been tied to were pulled out of the ground, sprinkling dirt and pumice around the campsite.

'I was wondering how long that would last,' Pita admitted.

After assessing the situation, he turned from the flapping sail and rifled through a pile of hardware salvaged from Manaia. Turnbuckles from the rigging caught his eye. The thought of driving beautiful pieces of machined stainless steel into the pumice felt sacrilegious, but he couldn't imagine what else they would do with turnbuckles, and the turnbuckles' long threaded-rod ends would make ideal tent stakes. Without further thought he began dismantling the turnbuckles and, after pounding their ends into the pumice walls of the gully, secured the edges of the genoa, confident the tent would survive a more serious blow.

Repitching the tent was a good distraction, taking nearly a half hour. Pita rinsed his hands, removed the burn cream from the first-aid kit and smoothed a healthy dab on Ani's shoulders. 'I'm sorry I got so mad. It was a great experiment, even if I don't like the answer.'

'You were right about one thing. The experiment was pretty nerdly,' Ani allowed proudly, earning a smile from Pita.

Pita returned to the issue of their longitude, 'Three-point-two thumbs east of Whangārei,' Pita recalled. 'Three-point-two days' sail east.'

'It would be at the equator. It's not quite as far here.'

'How far east would we be at this latitude?'

'2.75 thumbs north of Whangārei, we're at about thirty degrees south. That makes it pretty easy to calculate. Remember Kupe's triangles I drew in the sand?'

'Sure.'

Ani drew a circle representing the Earth on a page of her notebook. She drew the great navigator's triangles stretching from the centre of the Earth to 30 degrees south and 30 degrees north latitude. 'Each half of the vertical line is half as long as the diagonal lines. The question is, "How long is the horizontal line?"'

'Which would also be the radius of a slice through the Earth at our latitude.'

'Yes. Remember Pythagoras?'

'$A^2 + B^2 = C^2$'

'Right. The algebra in this one's pretty easy, but you don't have to follow along if you don't want to.'

'No, I'm the one who wanted to know how far it is. I want to follow along. Just go slowly.'

'Okay. Let's make A equal ½, C equal 1 and find B.'

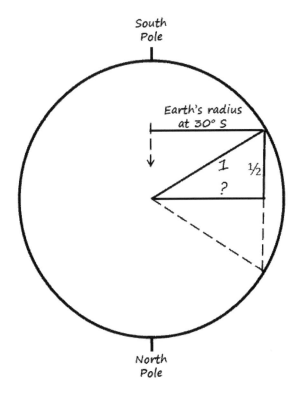

' $0.5^2 + B^2 = 1^2$

' $0.25 + B^2 = 1$ after we've squared the numbers – multiplied 0.5 times 0.5 and 1 times 1.

' $\qquad B^2 = .75$ after we've subtracted 0.25 from each side'
'Just a sec,' Pita requested as he caught up. 'Okay, go on.'

' $\qquad B = \sqrt{.75}$ as we take the square root of each side.

'I'm glad Koro kept a solar-powered calculator in a waterproof box. Calculating most square roots by hand gets pretty tedious.' Ani punched a few keys on the calculator and showed Pita 0.8660254 on the screen, 'That's the square root of .75. The number will actually go on and on forever, but 0.87 is fine for our purposes.'

'So, if the radius of the Earth at the equator is 1, the radius of a slice through the Earth at our latitude is 0.87.'

'Exactly. We don't have to get into pi and the real radius of the Earth, since we'd just be multiplying 1 and 0.87 by the same numbers. What matters is that, at our latitude, one thumb of east-west longitude is the same distance as 0.87 thumbs of east-west longitude at the equator. Or, since the Earth is a sphere, is the same as 0.87 thumbs of north-south latitude.'

'Yeah. So, if we're 3.22 thumbs of longitude east of Whangārei, instead of that being 3.22 days' sail, it's 0.87 times 3.22 days' sail.' Pita punched ' 3.22 x 0.87 = ' into the calculator and ' 2.8014 ' appeared on the screen. 'Two point eight days' sail east. How do we add those days to the southbound days?'

'As the crow flies, we could get a good approximation by using Pythagoras again. But we won't have another chance to estimate our east-west longitude, certainly not if we're afloat. I'm on my last page of sun data for Whangārei, my phone battery is dying and, anyway, once we're no longer on solid ground, we won't be able to determine solar noon accurately. We need all three of those things to determine our east-west longitude. We won't even have one.

'But we'll be able to estimate our north-south latitude from the stars any time we can see the night sky to the south. To sail back to Aotearoa, we'd want to sail south until we hit some safe latitude, say the latitude of the north end of the North Island, and then begin to turn south-westward. If we do that, even if the currents or winds carry us westward faster than we'd planned, we won't be likely to end up to the west of Aotearoa, crossing thousands of kilometres of the Tasman Sea on our way to Australia.'

'The tūpuna must have kept track of longitude by dead reckoning – knowing how far they'd gone and the direction they'd travelled,' Pita supposed. 'But I don't think they calculated in the same way we do.'

'They must have had the most elegant mathematical sense,' Ani volunteered. 'They could visualize the angles, combine south and west without drawing or calculating. Just sensing it. It's a combination of you and me – the way I calculate, and you feel. You just intuit how the wind interacts with sails, how the waves interact with the hull. There's so much more complexity in that; I don't know how to do that. Mathematicians like me are lazy in a way, finding number tricks to solve problems. But the mathematical elegance of our tūpuna, it's awe inspiring.'

'So, we're in the same situation our tūpuna were in when they voyaged from their ancestral homeland in central Polynesia to Aotearoa. We know how far east we are now. The tūpuna knew how far east they were from Kupe's sailing directions, which he learned from his intuitive dead reckoning while exploring. But, just like us, once they set sail, they had no way get a new fix on how far west they'd actually gone. Before they stood a chance of going too far west, they had to go south to a latitude where they knew a westward course would lead them to the long white cloud - Aotearoa. That's why the height of Māhutonga above the horizon is sacred to Karani.'

'If she only knew…'

'I learned the Pythagorean Theorem in school, of course. I trust my teachers, and you. But can you prove it to me? I'm the skipper,' he gave Ani a wink, 'I should understand everything under my command.'

Ani thought for a moment, took her notebook and began to draw. 'If you start with two identical squares, and you put four identical triangles in each, the leftover area in each must be the same, right?'

'Yeah.'

'Start with identical squares with sides equal to A plus B. If you put one ABC triangle in each corner of one square, the leftover area is C times C. On the other hand, if you nest two ABC triangles together just so in opposite corners of the other square, the leftover areas are A times A and B times B. $C^2 = A^2 + B^2$.'

Tūpuna Rock

The outer squares are the same area, (A + B) x (A + B). Four identical ABC triangles are in each square, one triangle in each corner of one square and two triangles together in opposite corners of the other square.

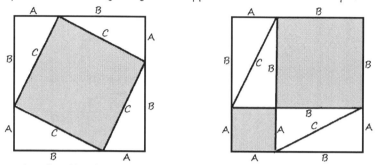

Since four identical triangles have been put inside each identical square, the leftover area in each square must be equal. $C^2 = A^2 + B^2$

'That's cool. Does it work for all triangles?'

'It only works for right triangles. But it doesn't matter if a right triangle is skinny or fat, it always works.'

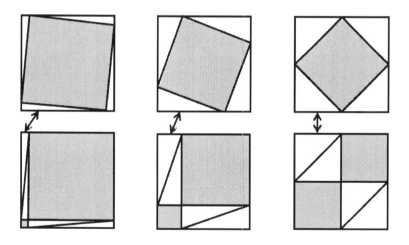

'What do you think skipper?'

'I love it. Thanks. Just for grins, how far would it be to Whangārei if we went straight?'

'We're 2.75 thumbs north and the equivalent of 2.80 thumbs east.' She took the calculator, and took the square root of the sum of 2.75 squared and 2.80 squared. '3.92 thumbs.'

'At 250 kilometres per thumb, that's almost a thousand kilometres. Nearly four days' sail with good winds in a good boat.'

'Yeah, using the wrecked floating dock as a raft is absolutely out of the question.'

'So, now that we've figured out where we are, what do we do?'

'This much is sure. If we just sit on the beach waiting to light a fire for a passing ship, we'll go crazy day by day. That could be months, maybe forever. We've gotta to keep doing things.'

'I read about a prisoner of war who kept his sanity by designing a whole house in his mind – board by board, nail by nail. He didn't have any tools or materials; he just designed it in his head. It kept him sane until he was liberated.'

'It doesn't have to be in our heads. We've got Koro's tools and driftwood on the beach. You've already started the smoker. We can make a more substantial shelter.'

Ani picked up her notebook with the thought of starting a list of things to build. She looked at the notebook's cover, 'Funny, I brought this along to write college admission essays. I don't think I'll get my UBC application in by January fifteenth.'

'When you do finally get it in, you'll have a heck of an essay to include.'

6   Tūpuna

A haku Ani caught in the morning was filleted, cut into strips, and placed in a pot to soak in a brine made of seawater and salt Pita collected from rocks splashed by the sea. In pans from Manaia's galley, they began to evaporate sea water to make brine to prepare additional fish for smoking.

Finally, rain arrived after a long dry spell, literally a real gully washer. Pita ran to cover the brine-drying pans and returned to find Ani standing out in the rain, letting the rain wash her peeling shoulders. 'You okay, sis?'

'I'm not sure how it'll feel when it dries, but rinsing the sweat and sea salt off sure feels nice now.

'Time to fill our water jugs,' Ani suggested, nodding toward the sail hanging low under the weight of rainwater over their campsite. Ani held Manaia's bailing bucket under the edge of the sail as Pita pushed the sail's bulging centre upward.

At the start, water poured over Ani as well as into the bucket, 'Sorry, sis.'

'No worries,' she replied with a laugh. 'It felt good, but let's try to not waste too much water.'

They filled all of their jerry-can water jugs and still the rain continued, prompting them to dump the first jug they had filled, which showed some silt from dust that had collected on the sail, and refill it. Finally, they simply stood in the rain, turning their faces upward, and revelled in the cool shower. When the rain finally stopped, they retired to the tent.

'So, we've got water, food and shelter. What's next?' Ani asked.

'We could build a proper shower.'

'I like that idea. I was starting to feel pretty salty.'

'The Haida made beautiful art,' Pita continued, proposing another activity but with no heart in it.

'They were in a land of bounty: salmon, deer, berries, running rivers and streams for water, wood for building and for cooking fires... I was beginning to worry that we'd run out of water before it rained; we might still. Once we've used all of the driftwood on the beach, we'll be tearing up the scrubby brush on this rock for cooking fires. That won't last forever.'

'So, we're not in British Columbia, the land of the Coast Salish People. And we're too far from Aotearoa to try to sail that piece of floating dock. What would our Māori tūpuna do, here?'

'They'd build a boat and sail to Aotearoa,' Ani answered without hesitation.

'With what? We've only got Koro's hand tools and no wood.'

'We've got those driftwood logs on the beach.'

'Carving logs into a sailable boat with hand tools would take months.'

'We've got months,' Ani observed before adding with a smile, 'We just won't have time to watch TV.'

Pita laughed. 'No, really, we don't even have a design. You can't just point to a log and imagine a boat we could sail across a thousand kilometres of open ocean.'

'Not just imagine, but we can design one.'

Pita walked the beach in the late afternoon, inventorying the available logs. There was a large log they noted on their first day, high on the beach, measuring ten metres long and seventy-five centimetres in diameter. He picked up a rock and hit it against the log all along its length. The big log sounded solid. A few other logs measured seven metres in length and up to fifty centimetres in diameter. One of those gave a dull thud when hit with the rock – rotten, but two of them sounded solid. Pita returned to the tent dragging a four-metre-long driftwood stick.

'We could build a ten-metre outrigger, or a seven-metre, twin-hulled boat.'

'Okay skipper, which would you prefer?'

'I'd rather face the ocean in a ten-metre boat with a seven-metre outrigger.'

'Let's design a ten-metre waka ama,' Ani concluded using the Māori term for an outrigger canoe. 'What's the stick for?'

'It's for a shower. Carving logs into a boat is gonna be a lot of hot, sweaty work. Every time it rains, we're gonna want to make the most of it.'

They tied the head of the midsize jib to a turnbuckle end driven into the ground near the end of their gully branch at the height of the plateau. The lower corners of the jib were tied near the ends of the stick, which was propped across the gully so the foot of the sail would hang about two metres above the ground. Rainwater falling onto the sail would run downhill and tumble off its foot.

Pita retrieved Manaia's boom and propped it across the gully about two metres away from the jib's foot. The boom conveniently rested between the gully walls at about head height. Unfurled from the boom, the intact, lower half of the mainsail dropped to provide a shower curtain. 'We're kinda in each other's faces all day long,' he explained. 'We should at least have a little privacy when we shower.'

'Thanks, bro.'

'Now, we just need a little rain.'

No rain was imminent, so they returned to talking about designing a waka.

'I don't know anything about any waka ama Māori ever built for sailing. All of the waka ama I've seen in Aotearoa are for paddling.'

'Does anyone build waka ama for sailing?' Ani asked.

'I know Hawaiians and Micronesians do. Micronesians have been building the best for centuries.'

'Tell me.'

'They build waka that don't tack to go toward the wind like other sailboats. They reverse direction to always keep the outrigger, the ama, to windward. That approach was actually pretty common all across Micronesia.'

'Why?'

'So the wind's pressure on the sail doesn't force the ama down into the sea. It also lessens the loads on the rig. The direction reversing manoeuvre is called 'shunting'. The kind of boat it called a 'proa'.'

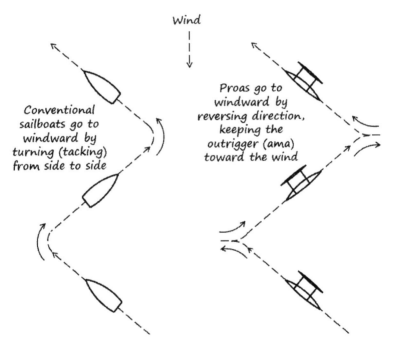

'Have you ever sailed a proa?'

'No,' he answered matter-of-factly, before adding, 'but it would be fun.'

'So, let's build a ten-metre proa.'

'I've never actually even seen one, never even seen a good drawing of one. Just read about them.'

'Then let's make a drawing. What do you know about them?'

Tūpuna Rock

'The hulls of the Micronesian proas are sleek. They're symmetrical from end to end, since they sail in both directions. But they're not symmetrical from side to side. The leeward side is flatter than the windward side, so the hull wants to turn away from the ama to compensate for the ama's drag.'

'Sounds elegant.'

'Yeah, it is.'

'So, let's design one.'

'But that's all I know about them, hulls that are sleek, symmetrical from end to end and asymmetrical from side to side.'

'Let's start there. What's sleek?'

'Smooth curves.'

'How asymmetrical?'

Pita silently visualized the drag of an ama and the flow of water around a hull, 'Maybe one third of the breadth is on the leeward side and two-thirds on the windward side.'

'Okay, let's start there.'

While Pita grilled fish and cooked fern root, Ani gave Pythagoras a workout on Koro's calculator.

Over dinner Ani shared a sketch. 'Is this a start for an outer shape? Ten metres long, 75 centimetres wide, two thirds of the breadth on one side of a line from bow to bow.'

'How do you do that?'

'Pythagoras again. First you calculate the radius of the curves you need, then you use that radius to draw the curve. The radii for this waka are so big that you couldn't just draw the curves by stretching a string from the centre. The radius of this curve,' she said pointing to the windward side of her sketch, 'is more than twenty-five metres. You'd be stretching strings across the width of a football pitch to draw the curves; the elasticity of the string would leave you with curves that are so wobbly that you couldn't use them.'

'How accurately can you and Pythagoras get them.'

'As accurately as you need. When you punch 25.25 into even a little hand-held calculator, it thinks 25.250000. How's to the nearest millimetre?'

'Plenty good. So, walk me through how you find the curve.'

'Imagine a triangle for each point along the curve. The radius is the diagonal line. The short leg is the distance from the middle of the boat to the place on the curve you want. The third leg is from the centre of the curve toward the end of the short leg.'

'Let me give you a couple of shippy terms,' Pita offered. 'The horizontal distance from the outer surface of the hull to a vertical plane from end to end is an "offset". Positions along the length of the boat are called "stations." You've seen the old movies when some old salt says, "Man your battle stations." There were cannons lined up along the length of the ship; he was telling everyone to go to their place along the length of the boat.'

'So, I was calculating the offset at each station.'

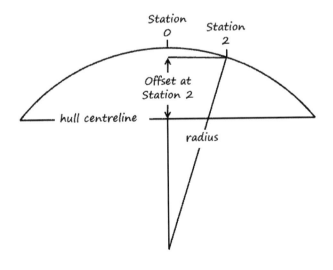

Tūpuna Rock

'Aye, matey, or I should say that in Māori, "Āe e hoa". Walk me through it. First, how do you find the radius?'

'We could call the middle of the boat "station zero". Start with a triangle with one leg as the length from bow to station zero. Since our waka is going to be ten metres long, that leg of the triangle is five metres long. We don't yet know the radius, the diagonal side of the triangle, or the third side. But we do know that the third side is the radius minus the offset at station zero. If the log is 75 centimetres wide in total, and two-thirds of the total breadth is on the side we're working on, the offset at station zero is 50 centimetres – a half metre. I guess it's not really the centre, since it's not symmetrical. Can we call that vertical plane from bow to bow the "reference plane"?'

'Makes sense.'

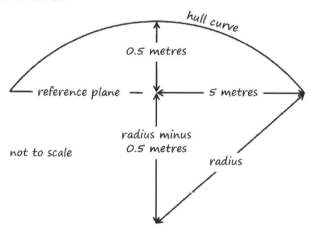

'Okay, "r" stands for "radius". By Pythagoras, $r^2 = 5^2 + (r-0.5)^2$. To get that r out of the last term, we use another math trick. $(A-B)^2 = A^2 - 2AB + B^2$.'

'Okay nerd, remember how I asked you to prove the Pythagorean Theorem? Show me how $(A-B)^2 = A^2 - 2AB + B^2$.

'Sure,' Ani assured as she began to draw in her notebook. '(A-B)² is the shaded area.'

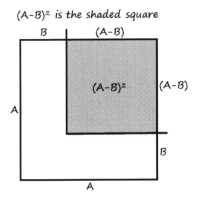

'Start with a square with sides equal to A. If you take away two strips that are A long and B wide you've isolated the square that is $(A-B)^2$.

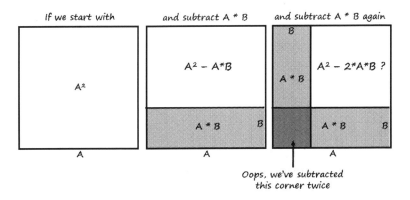

'But, you've taken that little square in the corner out twice, so you have to add it back in. It's B wide and B long, B^2. So $(A-B)^2 = A^2 - 2AB + B^2$.'

Tūpuna Rock

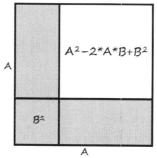

So we add the corner (B^2) back in.

$(A - B)^2 = A^2 - 2*A*B + B^2$

'Cool.'

'So our waka is five metres long from centre to bow and the offset at station zero is a half a metre.

'$r^2 = 5^2 + (r-0.5)^2$. By the $(A-B)^2$ trick, our formula for the radius can become

'$r^2 = 5^2 + r^2 - (2 * r * 0.5) + 0.5^2$.

'Wait a sec. I think I'm a picture guy. I like your graphic explanations, but I'm a little slower following the algebra.'

'You don't have to follow along if you trust me.'

'Oh, I trust you. But I'd like to follow, only slowly.'

'We've got time. We've got a whole lot of time,' Ani added with a smile.

After a pause Pita confirmed, 'Okay, I'm with you now.'

'Now watch this get easy!'

'$0 = 5^2 - (2 * r * 0.5) + 0.5^2$, after we subtract r^2 from both sides. Ani hesitated at each step until Pita nodded for her to go ahead.

'$0 = 5^2 - r + 0.5^2$, since $(2 * r * 0.5)$ is simply r.

'$r = 5^2 + 0.5^2$, after r is added to both sides.

'$r = 25 + 0.25 = 25.25$.

'The radius of that side of the hull is 25.25 metres. No square roots! We could do the whole thing in our heads! Good thing, cause the light's faded so much that the solar calculator isn't working any more. I'll calculate more curves tomorrow.'

'That isn't how the tūpuna did it.'

'No, it's not. But math is often only a system of tricks we've developed to describe the natural world. Our tūpuna didn't need tricks to describe the natural world, they understood it deeply, felt it intuitively. They had centuries of experience carving sailing waka, they experimented, and they knew what the fast ones looked like. They could see those hull shapes in their sleep; they didn't need Pythagoras. We've got only one ten-metre log – not much to experiment with – and we've never even seen a proa. I'm just using math tricks as substitutes for a small part of the knowledge we don't have.'

Ani hesitated for a moment before asking, 'Pita, I could create smooth curves from now until the end of days, but I can't tell which one is better than another for a waka. Sorry to have to dump it on you, but you're going to have to use your sailor brain, everything you learned sailing with Koro and listening now to the spirits of the tūpuna voyagers who visited this rock. You're going to have to choose the curves. Are you okay with that?'

Pita flipped through the sketches in Ani's notebook. 'I don't know how we're going to build it, but yeah, let's design it.'

~ ~ ~

Ani awoke to find Pita studying the hull outline she had created the night before.

'Sorry sis, it's not right.'

'Good.'

'How is that good?'

'It means you can sense what's not right. What did Edison say about inventing the lightbulb? "I have not failed; I've just found 10,000 ways that won't work." Each one got him closer to what worked.'

'But we've only got one maximum offset and one length. How many curves can there be?'

'More than Edison's 10,000. That's why your sailor brain is so important. What's your first impression of that shape?'

'It's too skinny at the ends, not enough buoyancy near the bows. A bow could bury itself into the bottom of a wave, flipping us over end for end. It's called "pitchpoling".'

'Good. I mean pitchpoling doesn't sound good, but the problem you described, tells me what to try next.'

'Oh, and we can't use the whole width of the log. That would mean the hull's height could only be half the log's diameter. Try keeping it to 60 centimetres wide.'

'Easy.'

While Ani punched away on the calculator, jotted notes and drew. Pita hiked down the gully to his smoker. Taking inspiration from pictures he recalled of Canadian First Nations smokehouses, Pita had assembled his small smoker from rocks stacked to make a chimney a metre-and-a-quarter tall and a third of a metre wide. He had left an opening at the base for feeding a fire.

Pita poked a stick through one end of each strip of brined haku and laid those sticks across the top of the smoker with the fish hanging down into its chimney. He covered the top of the smoker with doors from Manaia's galley cabinets to trap the smoke.

Pita started a small, smoky fire at the base of the smoker with the torch. Periodically, when the fire began to burn hot, he partially smothered it with a pan. When needed to sustain the fire, he added damp twigs to the growing pile of ash. It was a time-consuming process, but he had plenty of time.

After an hour or so Ani came to see him holding a notebook page with two hull shapes drawn on it.

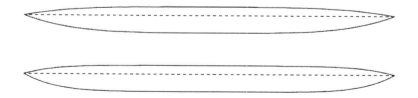

'They're better,' Pita said immediately, 'much less danger of pitchpoling. What are these curves?'

'I found the radii for a 25-station and a 125-station boat – our five stations squared and our five stations cubed. In the top shape each of our stations corresponds to its square on the 25-station boat. So for our station two I used the data for its station four (two squared), for our station three I used the data for its station nine (three squared), for our station four I used the data for its station 16 (four squared), finally, for our station five I used the data for its station 25 (five squared). In the bottom one, our stations correspond to their cubes on the 125-station boat, 8 for our 2, 27 for our 3, 64 for our 4 and 125 for our 5. You see what happens; the breadth of the hull is carried further toward the bows for more buoyancy at the ends.'

Using offsets from a longer hull at the square roots of their original stations, results in more breadth carried toward the bows.

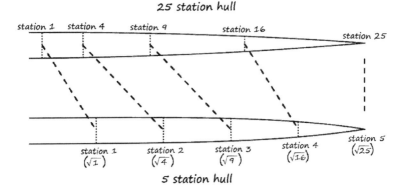

'I love the result, but I'm gonna have to study this drawing for a while to understand how you got there. Can I keep these with me while I tend the fire?'

Ani tore the drawings out of her notebook and handed them to Pita. 'I can do more if you use them to keep the fire going.'

'Oh, I won't do that.'

Pita studied the drawings as he fed dampened twigs into the smoking fire. He imagined how fast each hull shape would be and how it would behave in waves. He visualized the asymmetrical shape resisting the drag of an ama. 'Ani,' he asked, returning briefly to the tent, 'Can you draw a symmetrical ama? Use the full length of the seven-metre logs we've got and make it just 40 centimetres wide. Use the squared-station version.'

'Easy.'

Ani calculated some offsets, drew the ama Pita described and tore the page out of her notebook. The sun was climbing toward midday, so she wrapped her blue hikurere loosely over her shoulders before taking the drawing to Pita. 'I wanted you to have this right away.'

As Ani hiked back to the tent, Pita held the ama shape beside the hull shapes varying the distance between them. He was so absorbed in visualizing the interaction between the drag of the ama and the asymmetrical hull that he nearly let the fire go out. After he had the fire burning properly again, he returned to the tent. 'Have I worn out my welcome yet?'

'Not at all, this is easy for me, and a welcome distraction.'

'Can you make another hull shape with the squared-station shape on the windward side, but the simple curve on the leeward side? And put that ama you drew on the windward side so that the outer edges of the hull and ama are three-and-a-half metres apart. There are 12-foot deck boards on the floating dock we could use for the beams connecting the hull to the ama.'

'You bet.'

'You're the best. You know,' Pita added, 'we should use the right Māori terms for those beams and the main hull.'

'Do you know what they are?'

'Yeah, they're "kiato" and "hiwi".'

'Where did you learn that?'

'Remember last July when I went to that paddling regatta at Parua Bay?'

'Yeah.'

'I helped some guys rig their waka ama. They used the Māori terms. It was a great blending of cultures, sleek, high-tech fiberglass waka described in Māori. They called the hull the "hiwi" and the outrigger the "ama". Even Canadian sailors say "ama"; it's the same word in Hawaiian as in Māori. The guys at Parua Bay called the beams connecting the ama to the hiwi "kiato". I don't think there is an English word for those beams. Canadian sailors use the Hawaiian word, "iako". I've read that Hawaiians lost the "t" in their language, and it's often replaced with a "k". If you replace the Hawaiian "k" in "iako" with a Māori "t", do you see the similarity between "iato" and "kiato"? So "hiwi" was the only really new word of the three major parts of a waka ama: hiwi, ama and kiato.

'You know how nouns don't have a plural form in Māori? The guys at Parua Bay never added an "s" to "kiato" to make it plural. I asked them about it, and they said it just didn't sound right. They'd just say something like "that kiato" or "those kiato" to indicate what they meant.'

~ ~ ~

Ani was so excited by the new drawing that she had to take it to Pita straight away. Shoulders covered by her blue hikurere again, she hiked down to the smoker and handed it to Pita. 'I thought I'd draw in two kiato connecting the hiwi and ama.'

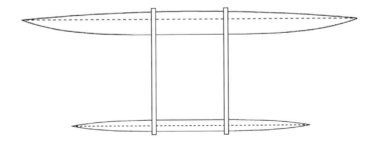

'Wow,' was all Pita could say at first. His eyes danced around the drawing, lingering longest on the bows of the hiwi. 'If this isn't it, it's sure close. Of course, this is only the top view, the offsets at one waterline. But I'm beginning to visualize the whole waka. If we could build this, we could sail it to Aotearoa. Let me keep looking at this while I finish the smoking. That shouldn't take too much longer.

While he finished smoking the haku, Pita recalled as best he could how the Haida, First Nations people of the Pacific coast of Canada, further dried their smoked salmon for long-term storage. Recalling photographs and visits to the provincial museum in Victoria, he lashed a drying rack together from driftwood sticks and used four additional sticks to prop it up facing northward, toward the midday sun. When he thought the fish was well smoked, he hung four strips on the drying rack and carried the remaining two strips back to the tent.

Offering a piece of his first experiment in smoked fish to Ani, he asked what she thought.

'To be honest, it's good but not great. It's a little bitter. If I just had this as survival protein, I'd be really grateful for it. But I don't think we could make a living selling it in a posh shop on Granville Island. Sorry.'

'No, it's okay. I was thinking the same thing. I used whole twigs this time, leaves and all. I thought the moisture in the leaves might help, but maybe there are tannins in the leaves that made it bitter. I'll try stripping the leaves off and just use the woody bits next time. Too bad there are no alder trees growing on this rock; they're a First Nations favourite for smoking. Still, there are a few different kinds of shrubs here. There's a lot to experiment with. But is this one okay for dinner tonight?'

'Absolutely.'

They enjoyed the good-but-not-great smoked haku for dinner, washing it down with healthy swigs from their renewed water supply.

'We've still got to sort out the front and side views of the waka,' Ani noted, 'but you think we've made good start?'

'I'd say, a "great" start. Boy, what a great day!'

Klaus Brauer

~ ~ ~

'I really want to flesh out that drawing you made yesterday. Are you okay making drawings based on nothing more than your little brother's gut?'

'If you mean the work of calculating and plotting, absolutely. I don't know about you, but I could go crazy just sitting on this rock. I enjoy math puzzles; they're the perfect thing to keep me occupied.

'Pita, you're closer to the tūpuna than I am,' Ani explained. 'You feel things they felt. I'm just happy if my math tricks can convert your gut feelings into a waka we can sail back to Aotearoa. So, yeah, I'm really okay following your gut instinct.'

'For sure?'

'For sure.'

'Okay, that front view of the hiwi. It's got to resist leeway, being pushed sideways in the water by force of the wind on the sail. So I think the leeward side should be really flat. As for the windward side, I can only project what I see in sails. You don't want it too flat, lest the flow of the water stalls – goes turbulent – and doesn't resist sideways motion as well. Sorry, that's not very specific.'

'No, that's a great start. Remember, I like math games. Iterating is fun.'

The leeward side was easy, Ani drew one using only the slightest curve. Her third iteration of the windward side made Pita smile, 'That's it.'

'What do you shippies call heights in a boat?'

'Waterlines.'

'That makes sense. Where to they start?'

'It's kind of arbitrary, whatever is convenient. Often, it's at the lowest point of the hull.'

'How about if our waterline zero is at the lowest point and the waterlines are 150 millimetres apart?

'You're the mathematician, if it works for you, it sure works for me.'

Tūpuna Rock

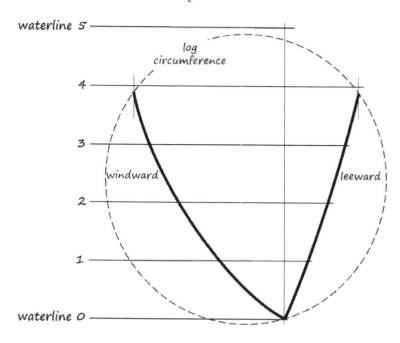

'Okay, now we know how broad the hiwi is at each waterline at station zero. Now all we have to do is figure out how long it is at each waterline. Basically, that's the shape of the bows.'

'That's harder to describe. Nearest the stem, the very front, the bows should be nearly vertical at the top, then they become more sloped and the curve gets tighter and then it progressively approaches horizontal.'

'Out of curiosity, what's the shape you're seeing in your mind?'

'It's a bow wave, when you're pushing it hard. Can you visualize it?'

'Not the bow wave. Let me go from your words. I think I know the kind of curve you described. Two questions. What's the slope of the top 15 centimetres?'

Pita held up Koro's folding rule, tilting it slightly to visualize slope of the top of the bows. 'In the top fifteen centimetres it only slopes back one-and-a-half to two centimetres.'

'Okay, second question: Down at waterline one, four-fifths of the way to the bottom, how far back are the bows?'

Pita borrowed Ani's notebook to sketch. After drawing and erasing and drawing again and again, he answered, 'About a quarter of a metre back.'

Ani worked for a few minutes, alternately punching keys on the calculator and plotting in her notebook. Finally, she showed Pita two curves.

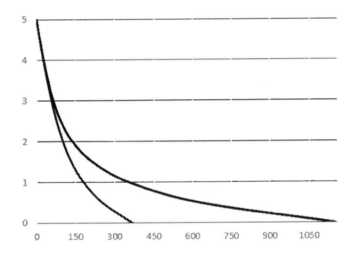

Pita studied the curves. 'Well, you've got it bracketed. The more upright curve is pushing the bow wave way too hard for the boat we want. The more relaxed one isn't pushing quite hard enough. Sorry, I can't be more scientific. It's just how I feel. The more relaxed one is close. It's just not pushing quite hard enough.'

Ani returned to punching keys on the calculator and plotting. When she showed Pita the results his eyes lit up.

Tūpuna Rock

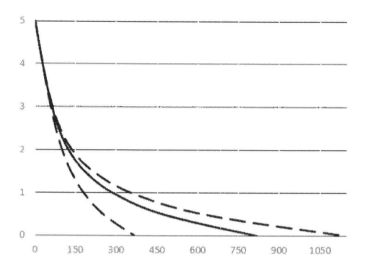

'The solid line's the one! What is that? You said you thought you knew the kind of curve I was describing.'

'They're exponential curves. That may sound mystical, but they're really simple. I had to reverse the order of our waterlines – I used zero for five and five for zero – otherwise they're just exponential curves added to that original slope you asked for.'

'Okay, but what are exponential curves in this case?'

'They're just a number taken to the power of the waterline. As examples, with my reversed waterline numbers, the station for the second waterline down was some number squared. It was that same number cubed for the third waterline down, and so on.'

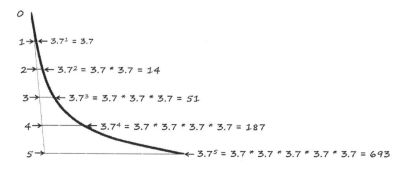

'How did you find a number to use so quickly?'

'You wanted the station of waterline one to be about a quarter of a metre back. Waterline one was the fourth waterline down. Some number to fourth power, the number times itself four times, was the key. When added to the original slope, three to the fourth power was less than 250 millimetres and four to the fourth was more than 250. Those were the first curves I showed you. Your reaction told me the base number should be closer to four than three. From your words I thought base 3.7 was about right. That's the last curve. Added to the original slope you wanted, the station at waterline one ended up at 257 millimetres – 70 millimetres from the original slope and 187 from 3.7 to the fourth power. That's almost exactly the quarter of a metre you visualized.

'It's really easy to calculate exponential curves, even on a basic calculator like this one. You start with 3.7, multiply that by 3.7 and write the answer down, then multiply that by 3.7 and write the answer down and so on.'

'Funny, the curve looks so natural, like a wave or something else found in nature.'

'Remember what I said? Math is mostly about describing the natural world. Really good mathematicians, better ones than I am, don't think so much about multiplying this times that. Actually calculating is kind of a trivial detail.

'You see a relationship between water and a boat moving through it in the curve of a bow wave. Mathematicians use formulas as shorthand for describing curves. The great ones visualize the relationship, the curve described by a formula, from the formula itself – no need to bother with calculating and plotting. For great mathematicians, physicists or whatever, visualizing relationships is all it takes to understand a problem. Our tūpuna were at that level, visualizing curves and carving waka with no need to multiply this times that. It's amazing to me that they could perfect the science and pass it on without the language of mathematics.'

Pita looked around the sail-covered gully and explained wistfully, 'I wish I could go back and be there in a waka shed when our tūpuna were building the waka for the migration to Aotearoa – just to watch and listen as the masters, the tohunga, taught young carvers.'

Ani formed the image in her mind, Pita sitting cross-legged on the sandy floor of a waka shed watching the gestures and listening to the ancient Polynesian language of the tūpuna. 'You would have understood it all, Pita.' In time she added, 'But sorry bro, you're stuck with me and Koro's calculator.'

Pita smiled, 'We're not stuck.'

Ani returned the smile. 'Now I know how long and how broad the waka is at every waterline. I can calculate the radius for each side at each waterline. Then I can calculate the offsets at each station at each waterline – the whole shape of the hiwi.'

'You've got a lot of number crunching to do. I think I'd better go down to the beach, look at the logs again and start figuring out how we're going to carve a waka.'

On his way to the beach, Pita passed by the smoker and his Haida-inspired fish drying rack. The strips of smoked haku he had left on the rack to dry and been pecked to bits by the island's birds. He had no idea why that didn't happen to the First Nations people in Canada, but saw that he needed to improve the process. The thought of another building project made him smile.

An acrylic foredeck hatch, the only transparent material at hand, was among the bits salvaged from Manaia. Pita built a drying box, like a greenhouse, using salvaged cabinetry as the sides and the foredeck hatch as the top. Driftwood twigs lashed together formed a grill to support the fish. After smoking the next batch of fish, Pita laid the fish on the driftwood grill and covered the box with the acrylic hatch. Laying the fish on a more horizontal grill actually seemed better than hanging it in the Haidi style. The sun never rose very high above the horizon back in Canada, so hanging let the sun's rays hit the fish more directly. On the other hand, the sun passed more nearly overhead on the rock, shining directly down onto the horizontal grill. The dryer worked. The sun shining through the hatch dried the fish while frustrating the birds.

7   Toki

Pita returned from the beach with a hopeful look on his face. 'Ani, I have an idea for transferring the offsets to the log.' He continued before Ani could respond, 'We have two doors from Manaia's galley cabinets. They're longer than the diameter of the ten-metre log. If we screw them to opposite ends of the log with their edges parallel to the reference plane and stretch fishing line between them at each waterline, we'll have waterline and offset references over the whole length of the log.'

Ani offered her notebook and pen, 'Can you show me?'

Pita made a quick sketch before continuing.

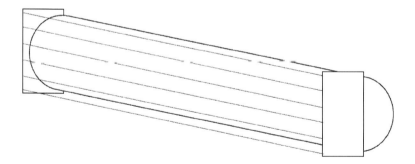

'The edge of the doors the fishing line is stretched between can be a known distance from the reference plane. If we subtract the offset at a location from the distance between the fishing lines and the reference plane, we know how deep to drill to mark the outer surface of the hiwi we want. We mark the drill at the right length with a piece of tape. When we've drilled down so that the tape gets to the fishing line, we stop drilling. We do that along each waterline at enough stations and fractions of stations to describe the whole shape.'

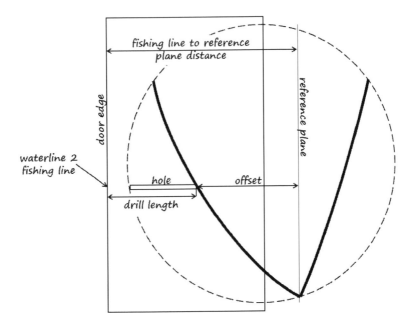

Ani took it all in, delighted by her brother's idea for transferring the numbers into the beginnings of a build plan. 'Wow. Do we have the drills we need to do that?'

'Old fashioned bloke that he is, Koro has a brace and whole bunch of bits in his tool box. He even has some ship augers for the really deep holes.'

'Any ideas for how to proceed from there?'

'If we saw across the log down to near the bottom of the holes every centimetre or two, we should be able to break out chunks to get near the final hiwi shape. Beyond that it's adze and chisel work.'

'Do we have an adze?'

'No. But do you remember the American framing hammer dad gave Koro?'

'The one with the straight-ish claws on the back?'

'That's the one. If we sharpen the underside of the claws, we can use it as an adze.'

'If we turn Koro's hammer into a Māori tool, we'd better call it by its Māori name.'

Pita smiled, 'Yeah, that'd be "toki".'

'How do you know that?'

'One of the voyaging organizations in Aotearoa is called the "Te Toki Voyaging Trust" honouring the tūpuna, our tūpuna, who carved waka with adzes.

'Using the sharpened hammer as a toki won't leave a very smooth surface; we'll have to finish up with a chisel, plane and sanding block.'

'Sanding block?'

'This whole rock is one big sanding block. I'm sure we can find flat pieces of pumice to use to sand the hiwi smooth.'

Pita hesitated before asking, 'So what do you think? Not about using pumice to sand, but about the whole idea of the doors and fishing line?'

'I think it's brilliant. It'll work.'

'There's more to figure out – rolling the log over to work the other side before we've gotten most of the weight out, hollowing it out, flotation – I could go on.'

'We'll figure it all out. I'm sure of it.'

'Okay, I'm gonna grab the galley cabinet doors, some screws and tools and head back to the beach.'

Pita returned to the log while Ani returned to the calculator and her notebook. After an hour or more, a moving patch of bright orange flashed into the corner of Pita's eye.

'I guess I can't sneak up on you while I'm wearing this. My shorts were wearing out from sitting on the rocks. I thought I'd make a skirt, a rāpaki, out of a piece of the storm jib. I figured this shredded section wasn't of any other use.' Ani did a slow pirouette causing strips of the shredded fabric to swing outward.

'You sure did find a shredded section. It looks more like a flax skirt, like a piupiu, than a cloth rāpaki. I do think the orange works with the blue hikurere on your shoulders. All in all, it's very fashionable.'

'It's what all the wāhine on this rock are wearing this season,' Ani laughed.

'Actually, I came down to show you what I've got so far and see how you're doing. I've calculated some offsets at each waterline. The hiwi is looking like this,' she explained holding out a page from her notebook.

'Beautiful. But you couldn't be done yet.'

'Oh, no. I just did the offsets at each one metre station to do this.' Ani offered a look at another page of her notebook showing the offsets she had calculated.

Windward Station	0	1	2	3	4	5
			Offsets (mm)			
waterline 4	413	412	402	358	240	-10
waterline 3	356	355	346	307	203	-19
waterline 2	276	276	268	236	151	-30
waterline 1	169	169	164	141	81	-45

Leeward Station	0	1	2	3	4	5
			Offsets (mm)			
waterline 4	200	192	168	127	71	-2
waterline 3	163	156	136	103	56	-4
waterline 2	113	108	94	70	37	-6
waterline 1	63	60	52	37	18	-8

'That's a table of offsets,' Pita announced with surprise. 'That's pretty standard in naval architecture.'

'Is that what it's called? I just thought it was a good way to present the data. Anyway, this just has 48 offsets. If I do four stations per metre, 25 centimetres apart, there will be another 120 for the hiwi. There will be 45 more for the ama, if it's symmetrical.'

Pita looked at the one-metre stations he had marked on the log, '25 centimetre spacing should be okay. I wouldn't want the spacing to be any bigger.'

'That's what I thought. Calculating more is easy. Should I worry about the drills and augers getting too dull to cut.'

'We'll have to learn how to sharpen them by hand with a file or stone. Just the same, I think it's right to start with a 25-centimetre spacing.'

Ani studied the doors fastened to the ends of the ten-metre log at an angle.

'The log has a little curve in it,' Pita explained. 'I thought we could use that curve to give us a little more height in the bows and a bit more depth amidships. The ocean, Mother Moana, left the log lying here with the plane of that curve about midway between upright and horizontal. I thought I could line it all up 45 degrees off horizontal. Koro's spirit level has a 45-degree vial, so that's as easy to do as upright or horizontal would be. It should even be easier to work the wood in that position. That line,' Pita pointed to a line drawn on the end of the log, 'marks the reference plane.'

'How far did you put the edges of the doors from the reference plane?'

'I can change it, but I made it 500 millimetres to start. I like keeping arithmetic simple.'

'It's easy to make mistakes. Keeping math simple is a good thing.'

Pita had already used the spirit level to mark waterlines and stations along the side of the log.

'Where do you think it would be best to start?'

'I think in the middle. I'm most concerned about a midships shape that resists leeway, resists getting blown sideways.'

Ani walked to the centre of the log and knelt at station zero. Checking her table of offsets and subtracting from 500 millimetres at waterline four she wrote 'drill length 87'. At waterline three she abbreviated 'drill length' to 'dl' and wrote, 'dl 144'. She continued for the remaining waterlines at station zero and stations one at either side of station zero. 'Pita, we've got stations of the same number on either side of station zero. What should we call them to tell them apart?'

Pita thought for a moment, 'We'll be sitting off the windward side to offset the force of the wind on the sail. Since we'll always be facing the hiwi from the windward side, let's call that station to the right as we face the hiwi from windward, "station one, right" and the one to the left as "station one, left".'

'That makes sense. "Sitting off the windward side"? I sure hope we won't be dragging our butts through the ocean all the way back to Aotearoa.'

Pita smiled, 'No, we'll build a platform to live on between the two kiato connecting the hiwi to the ama.'

Pita looked at the numbers Ani had written on the side of the log, put a piece of tape 87 millimetres from the point of an auger bit, put the bit in Koro's carpenter's brace and bored a hole at station zero, waterline four.

'May I do the next?'

'Sure.'

Ani checked her table and moved the tape's edge to 144 millimetres from the drill's point. Using the fishing line and the waterline Pita had drawn on the log to get the correct angle, she bored the hole at station zero, waterline three.

The two continued through the afternoon, alternating with the brace and bit, boring the remaining holes for all four waterlines at stations zero through three, right and left, stopping only because the sunlight was fading. Seeing the seven small piles of wood chips that had accumulated below the log where they had drilled gave them joy. They had begun carving their waka.

Tūpuna Rock

Another dinner of good-but-not-great smoked haku and rainwater might not have seemed fit for a celebratory feast, but it was.

The not-great smoked haku nudged them into a routine favouring fresh fish. Ani fished in the morning while Pita gathered fern roots. The fish were so plentiful and easy to catch that Ani suspected they were in a vast nature preserve where fishing was prohibited. Having no way to confirm her suspicion, she decided to keep reeling in fish to eat and, if necessary, ask for forgiveness later.

After breakfast, Ani crunched numbers in the tent for the remaining offsets while Pita bored holes, sawed down toward the bottom of the holes and pried out chunks from the log. Once Ani had finished calculating all of the offsets, she joined Pita in boring, sawing and prying. Each afternoon Ani hiked to the point with her fishing gear to fish for dinner.

One day, as a howling gale blew from the southeast, they hunkered down in the tent in the gully for most of the day. Toward late afternoon, after the storm had passed, Ani set off as usual toward the point to catch a fresh dinner. She returned in the day's fading light with a beautiful tāmure and a sad face.

The sad face made a far bigger impression on Pita than the fish. 'Ani, what is it.'

'Manaia is gone.'

Pita sat down heavily. 'I was afraid of that. The gale would have been blowing right into the cove at high tide.'

'I couldn't even see marks on the beach where she dragged anchor. She is just gone without a sign.'

'Ani, I always knew that would happen. We knew she would never sail again.'

Ani could see the tears forming in the corners of Pita's eyes. After a long silence she asked quietly, 'If you knew it would happen, didn't you want to save the anchor and chain for our waka?'

'I just couldn't leave Manaia on the beach without the anchor, even if it was just a gesture. It would be like giving up on her, and I couldn't do that. Does that sound dumb?'

'No, I get it.'

'I think we'll be fine with the little backup anchor.

'I had hoped someone, on an airplane or a ship, would see her hull lying on the beach and send someone to this rock to look for us.' He glanced upwards toward the genoa stretched overhead, 'We can still hope some airplane will see this tent, but mostly, we've got to keep carving our waka to sail ourselves back to Aotearoa.'

Ani put it bluntly, 'It's come down to either a huge stroke of luck or us saving ourselves.'

'We can save ourselves.'

Pita's confidence helped ease Ani from the melancholy the loss of Manaia had sent her into. 'You know Pita, I believe that, too.'

Ani sat beside Pita and wrapped her arms around him. 'I loved Manaia, too, Pita,' she acknowledged as tears skidded down their cheeks.

Pita wrapped his arms around his big sister and they mourned their loss together.

~ ~ ~

'I've heard the Canadian First Nations people burnt the centre out of their dugouts,' Ani began the following afternoon. 'Our hiwi is too narrow for that, but here's an idea. Our hiwi is a deep V shape. With sharpening, the augers are holding up pretty well. We could bore a line of holes parallel to the outside of both sides of the hiwi, use a chisel or something to knock out the wood between the holes and then pull most of the centre out of the log in one big chunk.' Ani made a sketch to show what she was thinking.

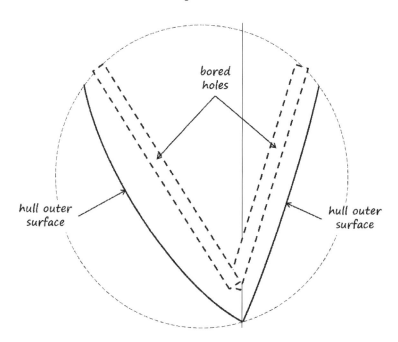

Pita studied the sketch and cautioned, 'We'd really have to watch the depths and angles.'

'I do numbers.'

'Yeah, you do. No worries there, you're the best.

'That centre wedge is going to be massive.'

'We could pull it out in pieces. Anyway, that's kinda the point. To get a lot of mass out of the waka.'

'What will we fill it with?' Pita wondered, 'If it just floods with water, we'd be better off just leaving the wood there.'

'You know the Japanese floating dock? I pass by it every time I go fishing. It has big foam blocks for flotation. We could cut those into triangular blocks that fit the cavity we've dug out of the waka.'

A huge smile grew on Pita's face. 'Wow, Ani! I was wondering how we were going to have a waka that didn't have the density of a log. I didn't really want us to sail back to Aotearoa on a couple of logs. I want us to sail back on a proper, fast waka. I want our tūpuna to be proud of us.' Pita looked down at the sketch and then squarely into Ani's eyes, 'From here on it's just work, a lot of hard work, but just work. I can see it all now. This is great, Ani. Our tūpuna will be proud of us.'

~ ~ ~

They were working on the windward side, Ani sawing down toward the bottoms of holes while Pita knocked out chunks of wood. Pita stopped, put down his improvised toki, picked up a straightedge and held it against the hiwi between stations one and two. 'Ani, something's wrong.'
'What?'
'Between waterlines two and four. It gets kinda flat going from station zero to station one. That didn't seem right, and now I see that it's becoming concave between stations one and two. The hole at station one and three quarters is deeper that the hole at station two. That can't be right, can it? The hole at station two is one of the holes we bored before, from that first set of offsets you calculated. Remember?'
'Yeah. Stop what you're doing. Something's definitely wrong.' Ani put down her tools and picked up the table of offsets she was working with.
Ani flipped back through her notebook to the first limited set of offsets she had calculated and compared them to the newer complete set. 'Crap!' In the new table she mentally compared the values for waterline three to the averages of those for waterlines two and four. 'Yeah, it is concave there. Damn, how did I do that?' She slumped, leaning over the log, studying the numbers and visualizing the digital curves. 'Crap, I'm sorry Pita. The new numbers for waterline three on the windward side are a simple curve, like the leeward side, they're not the squared station curve the windward side should have. 'Jeez, I'm sorry. How bad is it? How far have you gotten?'
'I don't think it's too bad. I've only gotten to station two, and I think the waka should float with the sea a bit below waterline three. We can just smooth out what I've done so far and carve

dowels from driftwood sticks to plug the unwanted holes. It'll be okay,' he assured.'

'I'm really sorry.'

'Ani, it'll be fine, honest. There are plenty of boats sailing around that aren't as fair as our waka will be. Let's call it a day, go eat some grilled fish and drink some vintage rainwater. If you catch something nice in the morning, I can spend the day smoking it. You should just chill until you're ready to calculate offsets again. You must be going cross-eyed from looking at numbers on that tiny calculator.'

'Thanks. Yeah, let's call it a day.'

~ ~ ~

As Ani returned from fishing the following morning, rainclouds were moving in. 'I'm first in the spa!' she announced excitedly.

'The spa?'

She pointed toward the shower, 'While you collect rainwater, I'm going in there.'

When the rain arrived, Ani shed her salty clothes into a bucket as it filled with rainwater. Not knowing how long the rain might last, she quickly stuck her head into the water tumbling off the foot of the jib, rinsing the salt out of her hair and then used her hands to wipe away the slick layer of water mixed with dust, sweat and salt that had formed on her skin. She turned toward the mainsail shower curtain, but was not ready to relinquish the shower. Looking up at the sky, into the falling rain, she leaned back into the stream of water cascading off the jib and became lost in the cool water flowing smoothly down her face, neck, torso, arms and legs. It was glorious.

In time Pita called, 'Hey, the water jugs are all full. Can I have a turn in the shower?'

'Yeah, just a sec,' she answered, roused from her reverie. She gave her storm jib piupiu and nylon hikurere a final rinse. After wrapping the dripping piupiu around her waist, she draped the wet hikurere over her shoulders and tied it in front. As she emerged from the 'spa' she smiled at her brother. 'You'll love it.'

~ ~ ~

Progress on the hiwi was slow but steady, in spite of the waterline three glitch. It took weeks to finish shaping the windward side of the hiwi to within a few millimetres of its final shape. It took weeks more to bore the holes around the central wedge. They decided to pull the centre out in three pieces: one between the stations where the two kiato joining the hiwi to the ama would be fastened, and the other two between the kiato stations and the bows. They left solid wood at the kiato stations to provide an opportunity to lash the kiato in place. Manaia's chainplates proved valuable as chisels for knocking out the wood between the holes bored around the central wedges. Actually pulling the wedges out took another week of chipping and coaxing before they came out, but they succeeded in pulling them out and smoothing the inside with the toki and chisels.

'Okay, we've gotten all the weight we can out of the log while it's lying on the leeward side, and it is still really heavy. Because it's sunk down into the sand, it doesn't even budge when we push on it as hard as we can to roll it.'

They both stepped back to study their half-carved, partly-buried hiwi and then at each other. 'Look at us!' Ani laughed. 'Do you think our tūpuna approve of our hats and sunglasses? I'm sure they're good with the rest.' Pita's sailing shorts had been in tatters by the time they pulled the central wedges out of the hiwi. To make a kilt-like rāpaki, he had cut open the bottom of a dark blue sail bag, tightening the drawstring at the top of the bag around his waist. With Pita in his rāpaki and Ani dressed in her piupiu and hikurere they had become a somewhat traditionally-dressed pair of Māori waka carvers – aside from Pita's floppy sailing hat and Ani's Vancouver Whitecaps baseball cap.

Tūpuna Rock

'I'm feeling the spirit of approval on the hats and sunglasses,' Pita answered, 'Maybe even a little envy about the sunglasses.'

Returning to the problem at hand, Ani proposed, 'If it were sitting on top of a couple of driftwood logs, we could roll it. How can we lift it up onto logs when we can't even roll it?'

'If it were on a boat trailer, we could lift the trailer's tongue.'

Ani got it. 'Because a trailer's wheels are near its centre of gravity, its balance point.' She thought for a moment. 'If we dig out under one end to near the hiwi's centre, we can tip it so that, as one end goes down into the hole we've dug, the other end goes up like a teeter totter. We can slip a log under the raised end just on the other side of the balance point. Then we can tip it back and slip a second log under.'

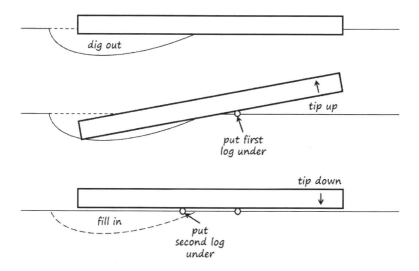

'Once we've got it up on two logs close to the hiwi's balance point, we can move it around by tilting one end up, moving that end in the direction we want to go, then tilting the other end up and moving it in the direction we want to go. It won't be fast, but it won't require more muscle than we've got.'

The idea worked like a charm. Manaia's backup anchor, a small claw, made an awkward yet effective shovel for digging under one end of the hiwi. Manaia's mainsheet block and tackle, hung from a driftwood tripod, gave them a five-to-one advantage for lifting the ends of the hiwi. Once the hiwi was up on the supporting logs, it just took one big heave to roll it over onto its windward side so they could begin working on the leeward.

They were back in business drilling holes, sawing and prying out chunks and then using their toki and chisels to approach the final shape. The process took two weeks, but it was a known procedure and proceeded smoothly. Along the way they dismantled the floating dock to salvage its lumber and flotation foam. Ani measured the central cavities in the hiwi and carefully carved blocks of the dock's flotation foam to fit the cavities. It took nearly three weeks of constant work to carve the ama by the same process.

One afternoon in mid-March, they finally 'walked' the hiwi and ama to a point just above the high tide line with a bow pointed toward the ocean. Exhausted from the effort, they retired to the gully for dinner.

'We've got a hiwi and an ama on the beach. What's next?' Ani asked.

'We need to have the kiato and the deck higher above the water than the height of the hiwi so waves don't hit them. I've been thinking we could have some smaller driftwood logs between the hiwi and kiato to raise the kiato and the deck up. We could notch the ends so they fit together at the corners like a log house. We could also use some big bolts from Manaia as spikes through the logs into the hiwi to be sure the logs don't shift.

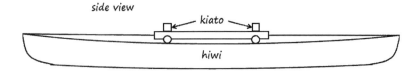

Tūpuna Rock

'Lashing the two kiato to the hiwi is a different story. The kiato have to be free to move up and down a bit while not moving fore and aft. The guys I helped at that paddling regatta last July taught me how to lash their waka ama together.'

'Do you remember how it was done?'

'Yeah. It was a really slick modern fiberglass waka, but they were fixated on lashing in the traditional way, even calling the process by its Māori name, "aukaha". It wasn't at all what I would have done if you'd just asked me to lash a kiato to a hiwi.'

Ani held out her notebook, 'Can you draw it?'

'Yeah. There's a beam in the hiwi under where each kiato goes. You start in the middle of that beam, then you go up over the kiato on both sides and out through holes in the sides of the hiwi. You stretch the line as it comes out of the hole and at each step.

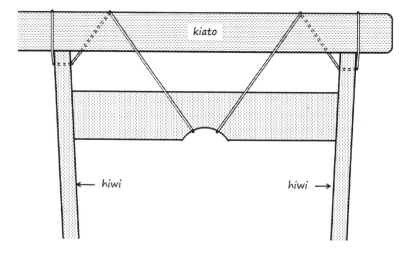

'Then, when the lines come out of the hole to the outside of the hiwi, instead of going up and over the kiato and back into the hiwi, you go under and around the kiato and then back under and then through a hole to the inside, stretching the line again.

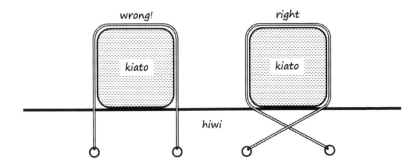

'Then you go back over the kiato to the centre of the beam. You repeat that again and again, each time stretching the line and overlapping the line you laid down in previous steps as you go over the top of the kiato. They say the overlapping stops the whole thing from coming undone if the line breaks somewhere.

'When you're down to a metre and half of line – I think they were using 30-metre-long lines – then you loop it again and again around the V you made in the centre, pulling it tight each time. That tightens the whole thing up.'

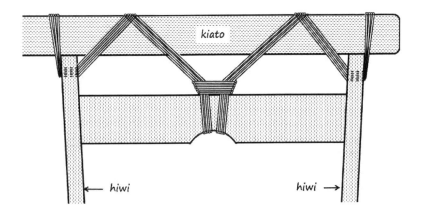

Ani studied Pita's drawings. She was perplexed at first, then it began to make sense, finally she smiled and shook her head, 'This is brilliant! They said this is the traditional way?'

'To hear the guys at Parua Bay tell it, Kupe himself used exactly this kind of aukaha on his waka centuries ago. With all

Tūpuna Rock

of their modern fiberglass, slick gelcoat and laminated kiato, they said there was no better way.'

'I can believe that's true.'

'What do you see in it?'

'First of all,' Ani offered, 'the line inside the hiwi holds the kiato down to the hiwi, but the long lengths from the centre of the hiwi to the outside provides longer lengths of line to stretch than if you just lashed straight up and down. Under the same load, a longer length of line stretches more than a shorter one. So, this method allows the kiato to move up and down a little, so sudden big loads don't break anything, the aukaha just gives a little.

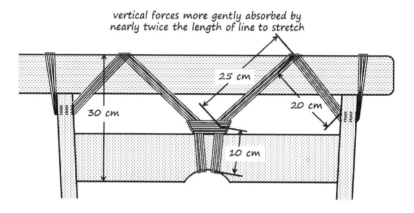

'Were they using a very elastic line?'

'It was nylon, about six millimetres in diameter. Of the modern lines, nylon is the most elastic. That's why you use it for anchor lines; its elasticity absorbs the shock of a boat bouncing around in waves while at anchor.'

'We don't have dozens of metres of six-millimetre nylon, do we?'

'No.' Pita paused to think. 'But we do have a hundred metres of nylon anchor line. I left the anchor chain with Manaia, but I fastened the chain directly to her bow. I saved the nylon line that normally goes between the chain and the boat.'

'That's much bigger than six millimetres in diameter, too big to wrap around the kiato more than a few times and way too stout for us to stretch.'

'Yeah, it's 16-millimetre line.' Pita thought more. 'But it's three-strand line, made from three smaller strands twisted together. Each of those three strands would be about the right diameter. We could try untwisting the anchor line into three separate strands.

'What's with the loop around the kiato on the outside of the hiwi? That doesn't seem like a good way to hold the kiato down.'

'It's not. Its purpose is to keep the kiato from moving fore and aft. If the line just went from the hole up and over the kiato and into the hole on the other side of the kiato, it wouldn't resist fore and aft movement much at all.'

Ani drew a dotted line on Pita's sketch of the wrong aukaha. 'Imagine a wave pushing the ama and kiato toward this dotted line. Is my dotted line much longer than your original vertical one?'

'It's hard to tell the difference.'

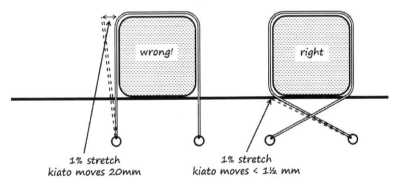

fore and aft motion resisted by line better aligned with forces

'Your vertical line, my dotted line and arrow between the endpoints is a triangle. Remember the ratio between those two legs of a right triangle?'

'Yup, cosine.'

'Right. Here's a little math-nerd trivia: Cosine of eight degrees, nearly four thumbs of angle, is 0.99. The difference in the lengths of those two sides is only one percent.'

Remembering that four thumbs of angle equalled one palm of angle, Pita held out his open hand. 'Wow. It's hard to visualize eight degrees, the number isn't intuitive, but a palm of angle is pretty easy to see. So, just one percent stretch in the line would let the kiato move through a whole palm of angle,' Pita concluded.

'But since the outboard lines in the proper aukaha are led fore and aft, they effectively resist fore and aft movement. With only a small angle between the line and the fore and aft force, one percent stretch wouldn't let the kiato move much more than one percent of the length of the line.'

'Yup. There's another thing bad about the wrong aukaha. The vertical line would have much more force on it. Pushing the centre of a 300-millimetre string eight degrees would move the centre 21 millimetres, but the overall length would only be 1% or 3 millimetres shorter. The end-to-end pull would be seven times more powerful than the push, 21 divided by 3.'

pushing the centre 21mm shortens the overall length by only 3mm
- the end to end pull is seven times greater than the push (21/3)

'Our tūpuna didn't need to know the cosine of one palm of angle. They would have observed the phenomenon every day. Try as hard as you can to hold a line straight when someone presses on the middle of the line with their pinkie. It's impossible to hold it straight. Think about a halyard on a mast. No matter how hard you haul on the line raising the sail, the wind can still blow the halyard around – the wind, not a lot of force on a skinny line!'

'Do you see how brilliant that whole system is?' Ani asked in rhetorical awe. 'The kiato is held down with enough elastic cushion to keep things from breaking while fore and aft movement is effectively resisted. That's all done with just one line; so over time, as the waka rocks through the swells and waves, the line can creep around and equalize the stresses so no part of the line is under any more stress than any other. Wow!'

With the principles of the traditional aukaha in mind, they lashed the two kiato between the ama and hiwi, and then lashed deck boards from the Japanese floating dock between the kiato. For shelter while underway, they added a lean-to deck house, a 'whare', on the ama side of the deck with its open side toward the hiwi – on the leeward side, away from the wind.

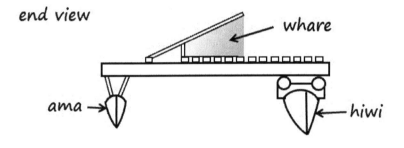

No longer a shower curtain, Manaia's lower mainsail, still attached to its boom, became the waka's mainsail. They cut it on a diagonal in the style of Polynesian sails, 'rā kautu' as they are called in Māori, and triple-sewed the cut end into a sleeve to slide over the lower section of Manaia's mast. The long seams required more thread than they found in Koro's sail repair kit, so they carefully picked seams out of the edges of the jib to salvage thread. Manaia's spinnaker pole was lashed along the leeward side of the deck. A line from the foot of the mast was looped around the spinnaker pole to guide the mast as it was moved fore and aft when shunting or trimming the sail. A second line from the foot of the mast was left to secure the mast

Tūpuna Rock

to the deck and carry the sail's loads once the mast was in position.

One end of the upper portion of Manaia's mast was lashed to the deck's edge out near the ama while the other end was lashed to the waka's mast as a strut to hold it upright side-to-side. The mast was held upright fore and aft by lines led from the bows to blocks near the top of the mast and down to Manaia's halyard winches, which were still mounted on the mast. The length of the lines to the bows could be varied using the halyard winches to move the mast when shunting or trimming.

With the waka moved to within a few metres of the ocean at high tide, they took the last few millimetres off the outer surfaces of the hiwi and ama with a plane and then smoothed the surface with blocks of pumice.

Pita and Ani stepped back to look at their creation. Ani asked, 'How does she look to you?'

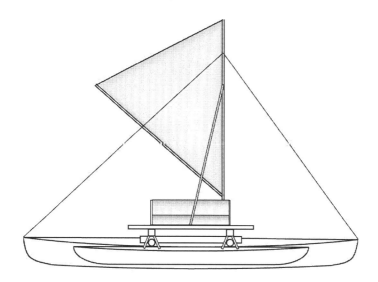

'Sometimes Polynesians call their waka "he". We don't have to follow all of the old traditions. Some traditions didn't accept women as navigators; boy did they miss a bet there. But what do you think about calling our waka, "he"?'

'I'm okay calling our waka "he", if,' she paused for dramatic effect, 'if you're okay with a female navigator.'

Pita chuckled, 'You can tell me where to go any time.'

'So, how does he look to you?'

'I focus on three things: what's in the water, what's in the wind and what's holding it all together – the hiwi and ama in the water, the sail in the wind, and the kiato, aukaha and other bits holding it together. Some of the bits out of the water and wind, like the logs between the hiwi and kiato, look pretty crude. But by my three key aspects, I think he's beautiful. How about you? What do you think?'

'I think we should make ready for a voyage.'

8   Making Ready

Ani needed a few clear nights to watch the stars from the observatory where she had a view of the horizon to the east and south. Pita slept nearby, wrapped in a sail, while Ani studied the stars and made notes all night. Ani slept late in the tent the following mornings while Pita smoothed the outer surfaces of the hiwi and ama with blocks of pumice. Near noon on one of those days, the bright orange of Ani's piupiu appeared to Pita a hundred metres down the beach. As she walked toward him, he was struck by her gait. He had long seen his big sister as a bit physically awkward, particularly since a teenage growth spurt that left her a nearly a head taller than he was. Standing with an arm resting on the hiwi, Pita watched Ani make her way up the rock-strewn beach. Her steps were fluid and sure-footed, not quite a dance, but genuinely graceful. She had walked that stretch of beach hundreds of times, to go fishing or to work on the waka, but her familiarity with the terrain didn't explain it. Connected by her tūpuna to Father Sky and Mother Earth, Ani was changed.

~ ~ ~

'It shouldn't be too hard to figure out what to take,' Pita observed with a laugh, 'We don't have much. But we're going to be underway 24 hours a day for maybe a week, with one of us both sailing and navigating while the other gets some rest. I want to be sure you're comfortable sailing our proa.'

'And I want to be sure you're comfortable navigating.'

'So, let's take turns teaching each other. Where do we start?'

'I've got to take more night-time sightings to get ready. I'd like your help with that. Why don't I start, so we're ready when we get a clear night?'

'Makes sense.'

Ani paused for a moment to organize her thoughts.

'North-south latitude is going to be just the way Karani taught us. Although it dips below the horizon here, in April, Marere-o-tonga reaches its lowest point in the middle of the night. Once we get far enough south, we won't have the problem we had when we first got here of Māhutonga being at its lowest point during daylight. And, we'll have plenty of moonlight to see the horizon in the coming weeks. All we'll need is a clear night, once we get far enough south to actually see Marere-o-tonga at its lowest point.

'On the other hand, remember, we won't have even one of the things we need to determine east-west longitude – a solid foundation for an observatory, data on the sun's position or a way to tell time. It won't be possible to actually know how far east or west we've gone until we reach Aotearoa. We can only estimate changes in east-west longitude by estimating how far we've gone in what direction.'

'Dead reckoning.'

'Yeah. Why is it called "dead" reckoning?'

'I don't know.'

'How about if I just reckon and skip the "dead" part.'

'You're the chief reckoner.'

'Reckoning how far we've gone through the water is going to be a matter of just counting the seconds it takes for a bubble or something floating in the water to go the ten metres from bow to stern, and then doing the math to convert that to metres per second or kilometres per day. Ani did a little mental math before explaining, 'One metre per second is 86.4 kilometres per day. If it takes four seconds for a bubble to go ten metres that's two-and-a-half metres per second – 216 kilometres per day.'

Tūpuna Rock

Pita shook his head and chuckled at Ani's ability to do calculations in her head. He could not wish for a better chief reckoner.

'As for direction, both for a heading to sail and for reckoning where we are, we've got heavenly bodies. From sunset to sunrise, we can use the stars in two ways.

'The celestial south pole is about midway between Marere-o-tonga and Māhutonga. Māhutonga, the Southern Cross, is easy to recognize. Marere-o-tonga is about three handspans away from Māhutonga more or less along the axis of the cross. They're pretty far apart, so what I do is hold out both hands with my thumbs and pinkies spread as far as I can, with one pinkie at Māhutonga and the other pinkie at Marere-o-tonga. That makes it easier to judge the point midway between the thumbs, the celestial south pole.

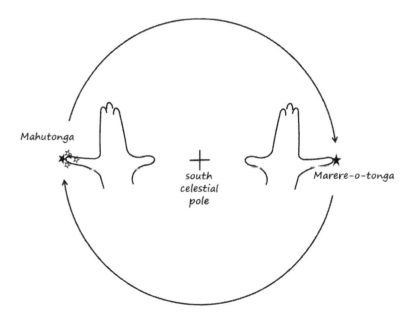

'The stars themselves will wheel around the south celestial pole, but the point midway between them won't move much, so we can use that point just the way we'd use the North Star back in Canada.'

Pita held his open hands out toward the southern sky. 'Got it.'

'The second way we can use stars is to give us a bearing to where they rise or set. Star compasses have been used by tūpuna all over Oceania for centuries, they show where stars rise and set. Instead of a magnetic compass needle pointing to the magnetic north pole, if you turn the star compass so the star's name is pointed toward the rising or setting of that star, the compass's south will point south, north to north and so on.'

'Like the floating card in a magnetic compass.'

'Pretty much.'

Ani unveiled a star compass drawn on a wooden plank.

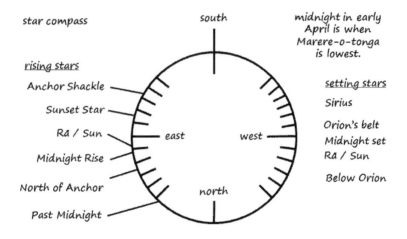

'You didn't have that in your dry bag.'

'No. This is from my nights studying rising stars at the observatory. Since I could only see the eastern horizon from the observatory, I've only got observations for rising stars. The tick marks on my star compass are palms of angle, four thumbs.

'On the next clear night, I'd like to go up onto the peak where we have a view of the horizon all around, confirm what I've got and note the positions of the setting stars.

'I don't know the proper name for most of the stars, so I made up names. There's a constellation that looks like an anchor with a bright star where the shackle would be; I call that star "Anchor Shackle". There's a star that rises at sunset; I call that one "Sunset Star". Not very imaginative, but they are self-explanatory.

'I've made our star compass specifically for early April. Sunrise and sunset move quite a lot over the course of the year, but not much at all over the course of a few days, so I just included the rising sun as another star. Some of my star names relate to when they rise or set. Those names will only be accurate for early April – by May my Sunset star will be pretty high in the sky by sunset.'

'No problem. We are going to sail in early April,' the skipper affirmed, 'and the self-explanatory names will help us find them.'

'Just the same, when we get back to Aotearoa, I want Karani and Koro to teach me the proper Māori names.'

Pita smiled. Ani was speaking in terms of 'when' they got back to Aotearoa, not 'if'. He was thinking in the same terms.

'And then there's the moon,' Ani continued, 'the moon is great because it is bright enough to see on a night that's too hazy to see stars, but the moon rises and sets in a different direction every night. The challenge is to predict where it will rise or set each night. I have data for moonrise and moonset for last December and January back in Whangārei.'

'More of your better-fishing-through-nerdliness data?' Pita poked good naturedly.

'Yeah, guilty. But I'm glad I've got it. Anyway, I plotted those two months of data to see what they would teach me. Each vertical line is a day. That thicker horizontal line through the centre is due east for moonrise and due west for moonset. The horizontal lines above and below the east-west line are spaced at one palm-of-angle intervals.'

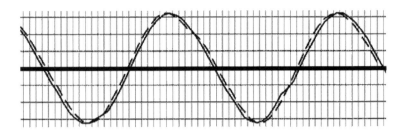

'I'm assuming your data for Whangārei don't have angles measured in palms and thumbs.'

'No.' she laughed, 'Those data are in degrees. But we don't have a sextant or any instrument that can measure degrees – just like our tūpuna. I converted the angles to something we could easily measure while we're sailing – maybe the same measure our tūpuna used. Anyway, the first line above the centre is one palm width south of east or west, the second line up is two palms south of east or west, and so on. I'm using up as south because we're going to be looking south throughout most of our voyage. It just seems more intuitive for our voyage – if you forget all of those maps created by non-Polynesians with northern hemisphere roots.

'The solid line is moonrises and the dashed line is moonsets. Here's what I see: Moonrise and moonset behave very similarly. Plenty close for our purpose. Second, there's clearly a pattern.'

'It looks like a sine wave.'

'Now you're sounding like the math nerd,' Ani gave a little chuckle. 'Generations of our tūpuna paid careful attention to the rising and setting moon. They didn't need to know a sine function; they knew the changes by heart. Most important, knowing the pattern deep in their soul, they knew what came next. If they knew where the moon rose for each of the past few days, they would know where it was going to rise for the next few days. We do a similar thing in mathematics. We fit the limited data we have to a curve we know.'

Tūpuna Rock

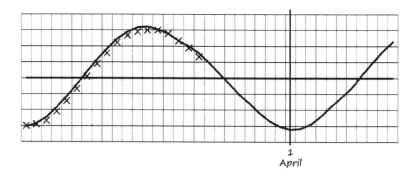

Ani showed Pita another plot. 'The solid line here is from mid-December to mid-January in Whangārei. In the evenings at the observatory, I was watching the moonrise as well as the rising stars. The Xs are my moonrise data starting about a month ago and fit to the December-January curve. My last data point is from March 23rd; after that, moonrise occurred during the day in bright daylight, and I couldn't see it. Moonset data could fill in a lot of the missing points, but, just like setting stars, I couldn't see the moon set from this end of the rock. If we go to the peak to map setting stars, I'd want to also plot moonset. If it fits on the curve, I'll feel good.'

'And if it doesn't?'

'We'll figure out why it didn't.

'Finally, there's the sun during the day. I don't know what the tūpuna had. Most likely they were so accustomed to the height of the sun at different times of day that they could just judge directions based on the height of the sun and its position. I'm not there – yet.

'I have an idea for a solar compass. It has a bent stick mimicking the arc the sun traverses in the sky and a second stick connecting the ends of the arc. When held at the proper angles with the sun arc shadowing the second stick, that second stick will be oriented east to west.'

'Held at the proper angles?' Pita wondered.

'The east – west stick must be level, and the sun arc must be tilted to the declination of the sun at noon, the angle between the noonday sun and the horizon. This time of year, the declination changes about one degree every three days. If we set the declination just before we leave, it'll be fine.'

'What can you use for a level?'

'Beach trash. I think one of those little plastic soy sauce bottles nearly full of water will work.

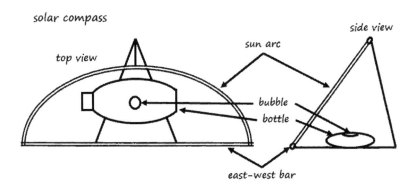

'So, if you hold the base level and pivot the whole thing until the sun arc shadows the east-west bar, the east-west bar will be oriented east to west. It's not perfect. It's really accurate in late morning and early afternoon, but it's off by a few thumbs at sunrise and sunset. I can use the height of the sun to estimate the correction I need to make.'

'Are you going to calculate some nerdly table of corrections?'

'Nope. I'm just going to visualize the sun's motion as it rises or sets, just the way our Tūpuna would. That'd be tougher back in Vancouver in summertime when the sun rises and sets at an acute angle way far north, but here, now, it'll work fine.

'That's the idea. What do you think?'

'Someday you'll have to show me the page of calculations in your notebook that's behind it.'

Ani thought for a moment and, surprising herself, admitted, 'I didn't do any calculations. I just visualized the sweep of the sun through Father Sky above the sphere of Mother Earth.'

Tūpuna Rock

' "Just visualized",' Pita echoed with a smile, 'you were listening to our tūpuna.

'It's really clever, Ani. And all just sticks, beach trash and, I assume, some fishing line to tie it all together.'

'That's all.

'Just two more things.

'We'll see some signs as we approach Aotearoa. One is, well, Aotearoa – the long white cloud our tūpuna navigators named the island for. From far away they could see clouds that form over mountains where the winds pushed air up to cooler altitudes.'

'How far away can you see the clouds?'

'That depends upon how high they are. Meteorology isn't my thing, so I've been trying to recollect the height of the clouds that form over Vancouver Island relative to the height of the mountains. There are a lot of 2,000 metre peaks on the island. Best I can estimate, 5,000 metres is pretty common for the cloud tops, 10,000 metres would be about the highest.'

Pita stopped to visualize clouds forming over Vancouver Island. 'How far away can you see a 5,000-metre-high cloud?'

'Our tūpuna knew it from experience, but our friend Pythagoras can help us recover what our tūpuna knew.'

Pita gave a laugh, 'I knew he could.'

'We know the circumference of the Earth is 40,000 kilometres. That means its diameter is 40,000 divided by pi, 12,732 kilometres. Earth's radius is half that, 6,366 kilometres.

'Here comes Pythagoras. We've got a triangle where A is the radius of Mother Earth and C is the radius of the Earth plus the height of the cloud. B is the distance from which you can see the cloud over the horizon.

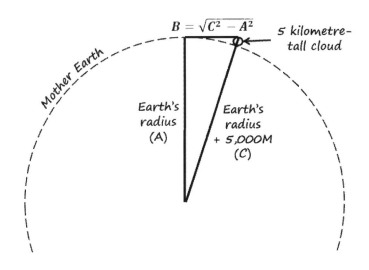

'The distance is the square root of 6,366 + 5 squared minus 6,366 squared. Okay, I'll admit I've got to use the calculator for that.' Ani punched buttons on the calculator for a few seconds, 'It's 252 kilometres.'

'About one day's sail.'

'Yeah, as it happens, about one day's sail.'

'What if the cloud is 10 kilometres tall?'

'Same approach.' After another few seconds Ani gave the answer, '357 kilometres.'

'That's not even half again as far.'

'Because the Earth curves away from us rather than dropping away at a constant slope.'

'So, when we first see the long white cloud, we're within a day or a day-and-a-half sailing of Aotearoa.'

'Yup.'

'So, we've got Marere-o-tonga for latitude, stars and the moon for bearings at night, the sun for bearings during the day and the long white cloud to tell us when we're getting close.'

'Yeah, just the things Kupe knew centuries ago, although he certainly knew these things better than I do.'

'Maybe he was actually a she,' Pita proposed, offering the thought not at all gratuitously.

Ani looked up and engaged Pita's eye, 'I like to think that's equally likely.'

'Anything else?'

'Birds. I don't know how far they go to feed. We can watch the birds that nest here as we leave and note how far they go. They'd give us a bearing to where they nest as they come out to feed and leave to return to nest at night. I don't think there's any place for birds to nest between here and Aotearoa, so we'll likely see the long white cloud before we see birds coming out to feed, but it's something to keep in mind.

'What do you think, Pita?'

'Our tūpuna found Aotearoa. We will, too.

'It's my turn to talk about sailing, but it's getting late. How about if I start in the morning?'

'Good idea.'

Wet or dry, their nights on the rock were all pleasantly warm. That particular night was overcast without threatening rain. Lying on Manaia's settee cushions, snuggled up in sailcloth, they both fell asleep shortly after nightfall, eager for the next day's learning and a clear night on the crater rim.

In the morning, Pita sketched in Ani's notebook and organized his thoughts, while Ani fished on the point. She returned in late morning with two āhuruhuru and a haku. While Ani grilled the āhuruhuru for brunch, Pita filleted the haku and put it in brine in preparation for smoking.

'Don't we already have enough smoked fish for a thousand-kilometre voyage?'

'Enough? Yeah, I think so. But can you have too much?'

'I guess not.'

As they ate grilled āhuruhuru and fern root, Pita began, 'Let me start with the basics. Sailboats are most efficient when the wind comes from the side, not from behind as you might think.'

'So, the east wind that we have so much of the time here isn't a bad thing.'

'Not at all. If the wind is directly behind you, it stands to reason you can't go faster than the wind. You go slower, in fact, partly because the wind doesn't flow smoothly around the sail. But look at this,' Pita pointed to a sketch.

'I like your idea of having south be up, so here's our waka headed south. The wind is coming from the east, just like most days have been here. With the sail well-trimmed, the wind flows around the sail like the airflow around an airplane's wing. The airflow is said to be "attached", sliding along both surfaces of the sail – not just on the windward side, but also on the leeward side, the downwind side. It's kinda like when you hold the back of a spoon in the stream of water from a faucet. The water sticks to the back of the spoon and flows off in the direction of the bottom edge of the spoon.

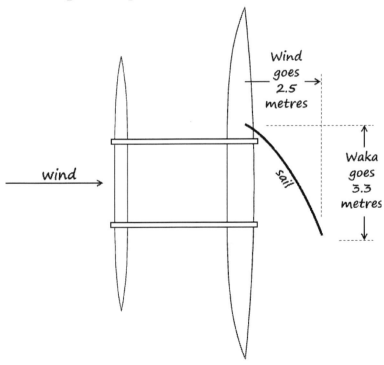

Tūpuna Rock

'The air on the windward side pushes on the sail, while the air flowing along the leeward side kinda sucks the sail along. It's a case where sucking is really good,' Pita added with a smile. 'As the boat gets going, that wing action becomes really effective and you have what I show here. As the wind goes two-and-a-half metres from side to side, the waka has to go three-and-a-third metres to get out of the wind's way. You can go faster than the wind.'

'Is that why you would often tack Manaia when you were going with the wind?'

'Yeah. When you're running – that's what sailing with the wind coming from directly behind you is called – you do have the wind pushing against the windward side of the sail, but the wind just tumbles around the edges of the sail and doesn't flow smoothly on the back side. You lose all of that energy you got from the back of the sail when the wind was coming from the side.'

'What you recall me doing was broad reaching – "reaching" is what sailing with the wind coming from the side is called. "Broad reaching" is reaching with the wind from midway between the side and the stern. If you really want to go to where the wind is headed, the trick is to get the wind coming from as far back as you can while still having the air flow smoothly on both sides of the sail. After a while, to get to your destination, you turn, "jibing" on a typical boat, to a broad reach with the wind coming from the other side. You zig-zag toward your destination. The distance travelled is longer, but you get to your destination faster.'

'Got it. What do you call the reach in your sketch, where the wind is coming directly from the side?'

'That's called a "beam reach".'

'What about going toward the wind? I understand the concept of tacking and shunting our proa, but how do you know what angle to steer toward the wind? What's too close? What's too far off?'

'You can feel it,' Pita answered without thinking.

'You can feel it,' Ani laughed. 'I can't.'

Pita laughed, 'Yeah, I guess that wasn't very helpful.' He thought for a while and then began to sketch. 'The maximum speed a sailboat can make depends upon the angle of the wind. It's different for different boats under different conditions, but the relationship usually looks something like this:'

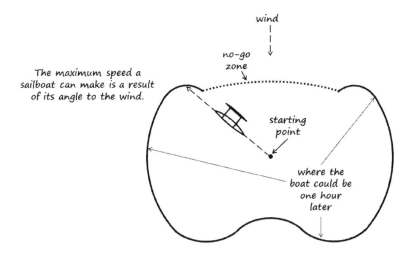

'I've seen drawings like this in books about sailing, but actually, the shape lives in my head from all of those days with Koro making a game of sailing for me. That solid line, the one that looks kinda like an apple, shows boat speed, how far a boat could go in an hour from the starting point. You can't sail directly into the wind or nearly into the wind; that's the "no-go zone" I indicated with a dotted line.

'I've always got the desired course in mind. So I just feel the boat's speed – the sound of the water flowing past the hull, the signs all around me – then I visualize that image and steer the course that takes me to the highest edge of the apple.'

Ani smiled, 'It does look a lot like an apple. It's a very organic shape.'

'I think it's one of the reasons I love sailing. No matter how modern the materials or the construction of our boats, to sail well you have to recognize your place in an organic world – your place inside that apple.

'There's a family of equations that have elements of your apple. They're called "cardioids".'

'Hearts?'

'Yeah, they resemble hearts so we call them by the Greek word for "heart".'

'I don't know if our tūpuna called the shape an apple or a heart, but I'm certain they knew it. It was so second nature to them they may have never even drawn it, but they felt it deep in their sailors' souls.'

'I don't think I could ever develop a single cardioid equation for your apple. You'll just have to teach me to feel it the way Koro taught you.'

'Okay, I will. Doesn't Pythagoras have anything to say?'

'Only about the easy part.' Ani sketched a triangle inside the apple Pita had drawn.

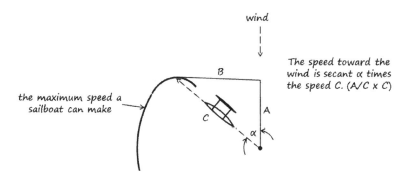

'If the wind is blowing from your destination, along A, you're finding the angle, alpha (α), that gives you the highest value of speed (C) times the secant of alpha. Secant is just A divided by C.'

'Is it okay if I like my way, our tūpuna's way, better?'

'Yeah, I like our tūpuna's way better, too,' Ani allowed. 'I may still visualize that triangle, but I need to develop our tūpuna's sense of the relationship between wind angle and boat speed.'

'You will.

'That basic understanding of sails applies to both Western and Oceanic sailing,' Pita explained, returning to his lesson plan.

'I think the most fundamental way Oceanic boats differ from Western boats is in how they steer. Western boats use a rudder. Rudders are effective, but they're inefficient. Think about it; a rudder is a board you turn sideways to the flow of the water. They create a lot of drag. Oceanic skippers steer by elegantly balancing the forces acting on a waka.

'The forces in balance are above and below the waterline: the wind's force on the sail and the force of the water on the hiwi and ama resisting the waka being pushed sideways by the wind.

'To think about the force on the sail, we talk about the "centre of effort" which is essentially the centre of the area of the sail. We usually determine that by drawing lines from each corner to the centre of opposite side. It's the fore and aft position we really care about, so I could probably just ask my sister where it is.'

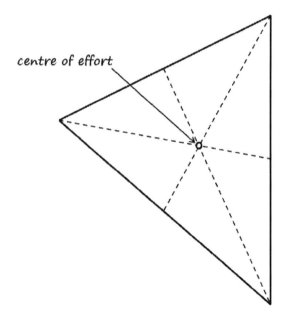

'For that triangle? It's one third of the way from the vertical edge to the opposite corner.'

'See, that's why I brought you along on this little adventure.'

Ani laughed, 'I wouldn't want my little brother out here all by himself.'

'Okay, that's the sail's centre of effort. The second bit is called the 'centre of lateral resistance'. It's a good name. A hiwi and ama resist being pushed sideways, laterally, in the water. That's why I wanted the leeward side of our hiwi to be so flat, to resist being pushed sideways. The centre of lateral resistance is just the middle of that resistance. If you think of a hiwi as a teetertotter on its side with various number of kids, or All Blacks rugby players, pushing on it, the centre of lateral resistance is the fulcrum, the pivot point.

'Since our hiwi is symmetrical from end to end, the hiwi's centre of lateral resistance is at station zero. The same is true of the ama. So, our centre of lateral resistance is generally on station zero close to the hiwi between the hiwi and the ama.'

'Generally?'

'Yeah, we could move it by putting more weight on one end of the waka than the other. That would put more of that end underwater generating more lateral resistance; the centre of lateral resistance would shift toward the heavy end. I don't want to plan on doing that. Sailors all around Oceania have been balancing the forces much more elegantly for centuries.

'First imagine we put our mast at the centre of our waka. The centre of effort of the sail would be aft of the centre of lateral resistance, pushing the bow toward the wind. That's called a "weather helm".'

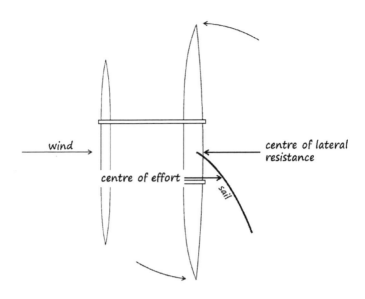

'Now imagine that we put the mast way far forward so the sail's centre of effort was forward of the hiwi's centre of lateral resistance. Then the wind would push the bow away from the wind.

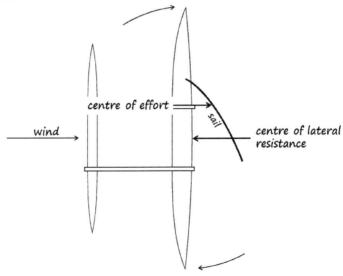

Tūpuna Rock

'Do you see the balancing act going on?'

'Yes, you manage the sail's centre of effort relative to the waka's centre of lateral resistance.'

'Right. Here's how the Ocean People learned to manage that. Start with the sail's centre of effort just a bit aft of the waka's centre of lateral resistance. The waka will have a subtle tendency to turn into the wind – a slight weather helm.

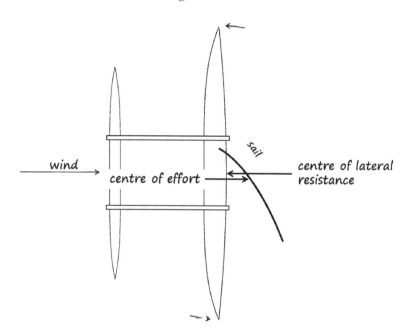

'Now you add a steering oar. But you don't turn the steering oar like a rudder creating the kind of drag a rudder does, you just dip it into the water in line with the direction of travel. Just like the hiwi, the steering oar has a centre of lateral resistance, resisting the waka being pushed sideways through the water. The steering oar isn't as big as the hiwi, but it's centre of lateral resistance is well aft of the hiwi's centre of lateral resistance. The average of those two centres of lateral resistance can be a bit aft of the sail's centre of effort, nudging the bow away from the wind.

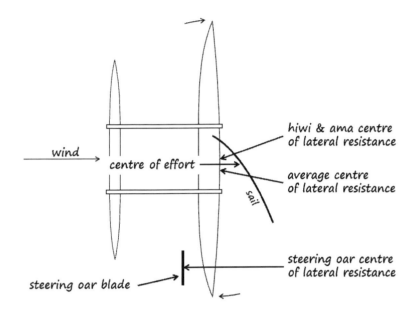

'You can dip the steering oar into the water every now and then, to steer a very subtle zig-zag course, or you can just dip part of the steering oar blade into the water to reduce its lateral resistance. That could put the average centre of lateral resistance in line with the sail's centre of effort, keeping the waka headed straight ahead.'

'That's beautiful! It's so elegant!'

'Yeah. With technology like that, people in wooden sailing canoes could settle islands scattered across the whole huge Pacific Ocean.'

'Our tūpuna.'

'Yeah. We've got our tūpuna's technology. We just need a sail plan.'

'Sounds like I'm back on.

'Once we get far enough south to see Marere-o-tonga at its lowest point, we'll be able to determine our north-south latitude accurately. On the other hand, like I said before, once we leave this rock, we'll have no way to confirm our east-west longitude;

we can only estimate it by reckoning. Our estimates of direction and speed won't be perfect, and we won't know how far east or west the currents have set us. With a wind out of the east blowing us to the west, the big danger for us is to end up so far west that we miss Aotearoa entirely and end up headed across thousands of kilometres of the Tasman Sea toward Australia.

'Here's my recommendation,' Ani offered, showing a chart sketched in her notebook. 'South is up,' she reminded. 'Even though Aotearoa is to the southwest, we start out heading due south until Marere-o-tonga is on the horizon at its lowest point. Then we can be certain of our latitude. Then we turn to the southwest. Even if we're way farther west than planned, we should still hit the coast of Aotearoa south of North Cape.

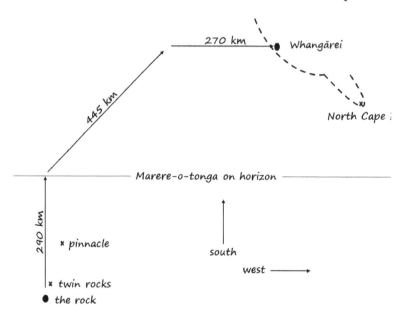

I wish my Aotearoa geography was better, or that I had coordinates for North Cape; I'm just going with my best estimate.'

'No worries.'

'Then we sail for nearly 500 kilometres on a south-westerly course, until we hit the latitude of Whangārei.'

'With Marere-o-tonga one-and-a-half thumbs above the horizon.'

'Yeah. Do you think Karani has any idea how important those nights on the beach have become?'

'She will soon.'

A smile spread across Ani's face, 'Yes, she will.

'Once on the latitude of Whangārei, we sail west until we see islands and coastal peaks we recognize. Then we sail to the beach in Smuggler's Bay and land our waka there.'

Pita smiled at Ani's detail. 'Do you have tide data for Smuggler's Bay?'

'My data are actually for Marsden Point, just across the channel.'

'I'm onboard with your sail plan. Out of curiosity, how much farther are we sailing by your plan than if we sailed straight toward Whangārei?'

'About two hundred kilometres, nearly an extra day's sail.'

'It's worth it to be sure we don't end up in the Tasman. Of course, if the wind holds out of the east, going straight would be a broad reach rather than the beam reach on that first leg. That would be faster.'

Ani interrupted, 'And more fun sailing.'

'Yeah, but later we can have fun sailing our waka fast. Now we should take the surer course back to Whangārei. It doesn't matter if we get there a day later. No one is expecting us.'

That no one was expecting them was a sobering thought.

~ ~ ~

The afternoon sky promised a clear night for mapping the setting stars. While Ani gathered what she needed to make the sightings, Pita gathered sails to bundle up in on the summit. Anticipating wind at the summit, to prevent the lightweight spinnaker cloth Ani slept in from blowing into an unruly heap,

Pita folded the light cloth into a blanket-size rectangle and stitched the layers together here and there to make a quilt.

Hiking across the rock for their astronomical campout, it seemed strange that, in their months on the rock, they had only been to the summit once before. On the plateau itself, they had rarely ranged more than a few hundred metres from their gully, usually simply to gather fern roots. Now, with their departure imminent, they hiked through the plateau's scrub and twittering birds with the sense of students who had studied in a new city for a year or two, and, on the eve of their departure, realized they had never adequately explored nearby neighbourhoods. It was difficult to call the scene pretty, but in some small way, it was theirs.

Pita set up camp beside Ani's observation location as she set up her 'apparatus' – a stake and an oar with loops of string tied around the shaft. Ready for the evening, as the sun declined, Pita took in the surrounding seascape. The specs of the twin rocks were no longer visible in the fading light, all he could see was a horizon of endless sea – three hundred and sixty degrees of endless sea. Months earlier, the sight had terrified him, but now, with confidence in their waka, a sense of direction and a sail plan, he was eager to set sail.

Having seen the sun rise two and two-thirds thumbs north of east from the observatory, Ani knew the sun would set at the same angle north of west. As the remaining stars set in turn, she compared their setting position to that of the sun. First the star she called 'Below Orion' set, then Orion's belt. 'Midnight Set' and Sirius set at nearly the same time in the middle of the night. Finally, she noted the moon setting three palm widths north of west, just as her plotted curve of the December-January data from Whangārei suggested it would.

Ani jotted her final notes by the fading moonlight before snuggling down into the bundle of sails near Pita. 'Get what we need?' he asked.

'Everything. I'm ready.'

'If we get enough sleep and the wind is right, let's give the waka a sea trial tomorrow.'
'Sleep well, bro.'
'You too, sis.'

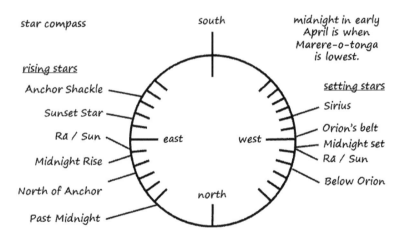

9 Sea Trial

'Ani, we should load everything we want for the voyage onboard for the sea trial. Then the waka will have the right weight and balance.'

'Makes sense.'

'There's something else.'

'Yeah?'

'It's a nice east wind we've got. But I've never sailed a proa.'

Ani smiled, 'Yeah, you mentioned that. I think this is the first proa you've actually ever seen.'

'Yeah, so I've got some learning to do. It could be that, while I'm learning, the wind pushes us so far west that we can't sail into the wind back to this rock.'

'Okay?'

'I am confident that I can sail our waka on a reach to the south in this wind. It's just getting back to the rock I'm not sure about.'

'Okay.'

'Are you really okay?'

'Yes, I'm really okay.'

'Then let's move the empty waka to the water at low tide and then load everything we need on board.'

'Āe, skipper.'

Ani looked around their campsite and caught sight of a large, white bundle. Manaia's jib no longer served as a shower, and she had wrapped it around the unused scraps of floatation foam from the wrecked dock. 'Can we leave the jib? I don't want the scraps of foam to blow off into the ocean.'

'We've got plenty of sailcloth left from the other sails.' Pita thought for a moment before adding, 'Mother Moana has been feeding us for months. We owe her our lives. The least we can do is to avoid fouling her. We can leave Koro's tool boxes on top of your bundle to be sure the whole thing doesn't get blown away in a gale.'

Pita opened the tool boxes and sifted through the worn saws, chisels, auger bits, brace and other tools. He removed the toki he had fashioned from a hammer to take along in case he needed to chop a rope in a hurry, left the remaining tools and latched the boxes shut. Koro's name and address were painted on the top of each. 'We can send someone back to collect these later. If we don't make it back to Aotearoa, someday someone will find these and know we've been here. The wear on the tools will tell them we built a waka to sail home.'

Ani's caution was piqued by Pita mentioning the potential of not making it back to Aotearoa. 'Are you worried about not making it back?'

'I'm not worried about finding our way. You've got that covered. And I feel really good about our waka. But the weather is always a risk. We don't have any weather forecast. We don't know what's waiting for us over the horizon,' he paused before concluding, 'and we both know just how bad it can get out there.'

'Do you still think we should go?'

'We've been lucky on this rock. If we stay, one day we could have an accident and get seriously hurt. We could get sick – appendicitis or something like that. Something important may be missing in our diet. Heck we don't even know how much vitamin C we've been getting. European explorers in the Pacific lost most of their crews to scurvy. Yeah, I worry about the weather at sea, but I think we should go. What do you think?'

'Let's learn how our waka sails.'

As the tide ebbed, they placed logs on the beach ahead of the waka to provide a sliding surface toward the water. Time after time they set the anchor in the tidal beach ahead of the waka and pulled the waka toward the retreating sea using Manaia's block and tackle for advantage. Each time the waka moved a metre or two toward the retreating sea they moved the log behind the waka that no longer supported it to the front, reset the anchor and pulled again.

When the tide stopped retreating, they began loading the waka: smoked fish, the barbeque grill with a full canister of propane, the sails they used as bedding – the genoa that no longer covered their gully was wrapped around the cushions to form a mattress. One twenty-litre jerry can full of water was enough for the two of them for the planned voyage, but they loaded three in case they were becalmed. As personal gear Ani took her notebook and fishing gear, as well as the solar compass she had devised. Pita's only tools were his rigging knife and the toki.

As Pita secured everything to the waka, Ani returned to scour their campsite in the gully one last time for anything they might need. She returned with nothing more than Manaia's tiller, which she had removed from Manaia before the sloop was washed away.

'What can we use that for?' Pita asked critically.

'This is what first taught you the touch of the tūpuna,' she answered holding out the tiller. 'That touch may be the most important thing we need to get us back to Aotearoa.'

Pita's look softened. 'Āe, I'll lash it to the deck.'

They set the anchor one last time as the tide began to flood and hauled the waka outward. As soon as he began to float, they climbed aboard, clipped their harness tethers to a line on the deck, hauled in the anchor and paddled northward to clear the north end of the rock. Then, it was learning time.

Ani sat quietly as she watched Pita struggle. The waka drifted aimlessly to the west. Pita asked calmly, 'Please move the foot of the mast to position two-right.'

'You don't have to say, "please".'

'That's just the kind of skipper I am,' he replied with a smile.

Ani moved the foot of the mast to the second position to the right, moving the sail's centre of effort, and repeated, 'Two-right.' The waka continued to drift aimlessly.

'Okay, let's try three-right,' Pita asked calmly as the waka continued to drift out of control. Reversing himself he asked for position one right. Ani moved the mast each time he asked. No matter the mast position, both the sail and the hull remained stalled. They were 'in irons' drifting sideways, westward, out of control.

'Two right should be the right position,' Pita explained uncomfortably. 'Let's go back to that.'

Ani moved the mast and watched Pita grow increasingly worried, silently pulling the sail in and then allowing it to run out, rowing with the steering oar to change the orientation of the hull to the wind. Nothing worked, the waka remined in irons, drifting westward. The thought that they had somehow fashioned an unsteerable raft, destined to drift out into the Tasman Sea toward Australia was inescapable. Pita fell silent, checking and rechecking the position numbers he had marked on the deck.

'Pita,' Ani began calmly, 'sit down on the back of the deck.'

Pita, frustrated and anxious, snapped at the suggestion, 'What? Is this a mutiny? Are you taking over as skipper?'

'No, damn it,' Ani stood her ground. 'Look at yourself. You've got your head down, studying numbers written on the deck. You're acting like a nerd, studying numbers to figure it out. We don't need another nerd on this waka. Only a sailor can figure it out. Sit on the back of the deck, watch the water at the stern. Hold the sheet in one hand, feel the wind's pressure through the sheet, just like you've been doing ever since you were a little kid. Hold the steering oar in your other hand. Let it tell you what the waka wants, just the way Manaia's tiller did.

Warming to Ani's idea, Pita offered, 'I think our tūpuna called the steering oar a "hoe".'

Tūpuna Rock

'Listen to them, Pita. Listen to our tūpuna.'

'I'll move the mast,' Ani explained, 'but don't look at the mast or the sail. Just watch the water at the stern, feel the pressure of the wind through the sheet and the water through the hoe. Feel the ocean through the waka and your sensitive butt.

'After you figure out what works, maybe we can use numbers to analyse why it worked. That'll help if we want to design another waka. But now we're adrift in the middle of the Pacific. This is the waka we've got, and you're the one who can figure out how to sail him.'

Pita understood and sat on the aft edge of the deck looking astern as instructed. He let the sheet run out as Ani muscled the mast from position to position. Each time the mast was secured in a new position, Ani said simply, 'Try this,' and Pita tried the sail by hauling in the sheet.

After perhaps a dozen frustrating tries, the eddy of the water spilling around the stern broke away from the stern as trace of a wake formed. Pita brightened up, 'Do more of the move you just made.'

Ani moved the mast as suggested. The sheet gave Pita a promising tug and the water flowing past the hoe spoke to Pita. 'Still more, please.' With that one more move, they began making way. They were sailing rather than drifting.

'Can I ask?' Pita wondered.

'Yeah, it's at seven right.'

Pita spun around to see for himself, 'That's not possible! With the mast that far forward, the sail should just be dragging us downwind.'

'Something about Western naval architecture must not apply to this boat, maybe not to all proas,' Ani proposed. 'Forget about what you expected and just learn to sail this waka.'

Pita learned quickly once he knew how to get the waka moving. Nodding toward the halyard winch on the left side of the mast, he asked, 'Let a few centimetres of line run around that winch to ease the masthead a bit forward. That'll shift the sail's centre of effort a tad more.' The waka settled down and really began to make way. With Pita tweaking the sail and managing the steering oar, the waka began to pick up speed northward. Pita heaved a relieved sigh, 'That was fun,' he muttered without meaning it.

'If you say so. We're sailing,' Ani added in relief.

'Yeah, we're sailing! We haven't been sailing for months. I've barely got it under control, but it feels good to be sailing again. Let me feel out this beam reach a bit more.' They sailed another kilometre northward as Pita requested subtle changes to the sail trim and managed the steering oar. 'I want to try sailing closer to the wind, to the northeast. We're gonna have to do that if we're gonna make it back to the rock. Move the mast back to six-right and be ready to adjust the masthead.'

'Six-right,' Ani repeated as she moved the mast and used a winch to pull the masthead aft.

Pita worked the steering oar to shift them to a north-easterly heading and pulled in the sail a bit. His face had been emotionless at best since they left the beach, but now a smile grew on his face, 'Ka pai! This is good, Ani.' He hesitated before admitting, 'I was really worried there for a while.'

'Yeah, I could tell.'

'Thanks for kicking me in the ass.'

'Any time you need it.'

They continued on the north-easterly heading for fifteen or twenty minutes, Pita getting the feel of the waka. 'Ani, are you ready to reverse direction on our first shunt?'

'There's a first time for everything,' she answered with a laugh. 'I'm ready if you are.'

'Okay, I'm going to turn due north with the ama toward the wind. When we come to a stop, I'll ask you to move the mast to six-left.'

'Got it.'

Pita worked the steering oar to turn them due north. As the waka came to a stop, he asked, 'Six-left, please.'

'Six-left.'

Tūpuna Rock

'Use the other winch to let the masthead forward, kinda the mirror image of what you had going before.'

Ani understood and began easing the masthead to the desired position as she asked playfully, 'Where's "forward" on a proa, a boat with two bows.'

Pita had to think about it before proposing, 'How about toward the "now bow", the bow pointing toward where we want to go.'

Ani liked the term "now bow" and confirmed, 'Toward the "now bow" it is,' as she finished positioning the masthead.

'That's good.' Pita trimmed the sail for the new heading and the waka reversed direction, heading off to the southeast. 'Wow, just like it's supposed to. Are you surprised?'

'I'm impressed, but I was prepared to be. I'm not surprised.'

Pita continued on the south-easterly heading for a few minutes. 'I want to go to a broad reach back toward the rock. Let's try the mast foot at eight-left.'

'Eight-left.'

'And the masthead forward a bit.'

'That's it.'

Ani studied Pita's face as they approached the rock. 'How does he feel?'

Pita was so pleased by the performance of the waka there was an element of surprise on his face as he looked across the deck at Ani, 'He feels good, very good.'

Ani noted that Pita was systematically looking at each lashing, rather than revelling in the sound of the water rushing past the hiwi and ama. His mind was an open book; he was inspecting the waka in preparation for a long sail. 'I didn't leave anything we need on shore, did you?'

'No, I didn't, and I like this wind. Should we just go?'

'Yeah, let's go,' she answered without hesitation. 'We've been missing for too long. Let's go.'

'Get ready to move the mast to seven-left.' We're going to turn south between the rock and the crag. 'Seven-left, please.

And haul the masthead back; I think you're getting to know the masthead position that goes with seven-left.'

'Yes. Seven-left.

'Pita, I've seen a small rock south-southwest of the crag while fishing. It's just above the surface at low tide, so it may be an underwater hazard now. Just continue on this heading until we're well clear of the rock.'

As they sailed between the rock and the crag, Pita called out a thanks and good bye to the ancestors, 'Ngā mihi, tūpuna. E noho rā.'

Ani echoed the thanks, 'Ngā mihi, tūpuna …' and then stopped herself, turning to Pita, 'The spirits of our tūpuna will be with us all the way back to Aotearoa. They will be with us forever.' After a moment's thought she added, 'And they will be on that rock forever, for whoever else may need to feel them one day.' Turning back to the rock, she called out, 'E noho rā, Tūpuna Rock.'

Pita smiled and echoed Ani's christening of the accidental home they were leaving behind, 'E noho rā, Tūpuna Rock.'

Pita steered clear of the reefs between Tūpuna Rock and the crag and the rock south-southeast of the crag, and then looked to Ani, 'Navigator?'

Ani looked back at the trail of bubbles in their wake to judge how much leeway they were making and then held her solar compass out level in the sun, turning it so the east-west bar was shadowed by the arc, 'Steer two thumbs further east. On that southerly heading we should pass to the east of the twins. There, that's good.' Ani watched the bow swing eastward, 'Hold this heading until we see Marere-o-tonga kiss the horizon.'

10   Voyagers

The sun was setting as the waka left Tūpuna Rock in his wake. Referring to Ani's star compass, the rising 'sunset star' provided a ready bearing. The wind was blowing steadily out of the east at a speed they estimated to be about 20 kilometres per hour. Although its sail was a bit on the small side, the waka sailed well in the available wind. A nearly full moon floated high in the eastern sky.

Watching bubbles float past in the moonlight, Ani calculated that they were making about three metres per second. 'I think we're making nearly eleven kilometres per hour.'

'I'll take that… …all the way back to Whangārei.' Pita was experimenting with fine tuning the sail's centre of effort by tilting the mast with the winches. He quickly got the hang of it and left the tethered steering oar lying on the deck. He steered into the night, with one hand adjusting the sail's angle toward the wind.

Pita remembered that sailors throughout Oceania often kept track of their course simply by feeling how a waka rocked in the swells. With the sail under control, he turned his focus to learning how to read the swells. He found them easier to read in the waka ama than in a conventional sloop and sat with his eyes closed for long periods, opening them periodically to check the stars. 'Ani, close your eyes and feel how the waka is rocking.'

'Okay.'

'Remember how this feels. The swells are coming from the east and we're headed south, so it's a very simple rock. The ama goes up, followed by the hiwi, then the ama goes down followed by the hiwi; there's no bow up and down motion. The direction of swells stays the same for days at a time. We should be rocking like this for as long as we head south.'

'I do feel it, but to be honest, I'd rather rely on the stars.'

'That's good until clouds move in.'

'Point taken.'

There was not a cloud in sight that evening. They could easily see the twin rocks illuminated by moonlight in the distance to the west as they passed.

Both voyagers felt a mix of elation and anxiety as they continued off across the dark sea. They were underway, headed toward safety, and their waka was sailing well; but they were leaving the relative safety of the rock behind. Ahead lay nearly one thousand kilometres of open ocean, an ocean they knew full well can turn angry and unforgiving.

~ ~ ~

'While we've got stars to steer by, I think it's my turn at the helm.' Ani proposed. 'Then you can get some rest. Can you trim the sail so I can steer by dipping the steering oar?'

Pita pulled the mast a little aft as Ani moved to the steering oar. Before she dipped the steering oar into the water, Ani watched the bow swing slowly toward the east. As she dipped the steering oar fully into the sea, the bow's swing stopped and then reversed to swing slowly toward the west. 'Cool!'

'You've got it, sis. I'm going to take you up on your offer and get some rest. The tradition among sailors is to take four-hour watches, but I've no idea how to tell when four hours are up.'

'We've got a big clock up ahead. Just imagine a line drawn between Māhutonga and Marere-o-tonga. That line is the hour hand on a twenty-four-hour clock. When it makes a sixth of a turn, I'll wake you.'

'Thanks.' Pita climbed into the lean-to whare deckhouse, wrapped up in the spinnaker quilt to stay warm in the night air

Tūpuna Rock

and closed his eyes. He did not, in fact, fall asleep straightway. As he lay awake with his eyes closed, he monitored the rock of the waka in the swells. Ani was doing a fine job of holding their course. In time the swells rocked him to sleep.

It was a beautiful night. Behind the waka, the moonlight reflected off the troughs of the smooth swells in straight north-south lines. Ahead, the sea was dark and the sky full of stars.

~ ~ ~

Ani woke Pita as planned. 'How did it go?' he asked.

'It was beautiful. Steering was easy. I had Māhutonga and Marere-o-tonga to guide me all night. Watching bubbles float by in the moonlight, I calculated we were making about ten kilometres per hour.'

Pita turned his head to face the breeze, 'In this wind?'.

'Yeah, it's been like this all night.'

'That's a good speed for this wind. Let's hope it stays just like this.'

Ani waited for Pita to get settled in at the helm, then crawled into the whare, wrapped up in the spinnaker quilt and promptly fell asleep.

Their second day at sea began as the preceding night had promised – a steady breeze driving the waka happily onward. After hours of trying a variety of lures and a long fight with a strong fish, Ani finally landed a mahi-mahi. After securing the fish onboard, Ani shook out her tired arms, 'I think "mahi" means "work" in Māori. If that's right, "mahi-mahi" is a good name for this fish. It's work-work to reel in.' The grilled 'work-work' was delicious. With fair winds, a sound waka and good food, the voyage was off to a great start.

~ ~ ~

'Ani,' Pita warned in the late afternoon. 'I see some storm clouds up ahead to the southeast. Could be trouble.'

'How much trouble?'

'Too soon to tell for sure, but I would like to avoid them. Our experience with storms out here isn't great. I don't want to turn due west while we're still so far north. How would you feel if I made our turn to the southwest now, before we've seen Marere-o-tonga kiss the horizon?'

Ani was torn. She wanted to be careful not to pass north of Aotearoa and out into the Tasman, but their experience in storms at sea had been terrifying. She studied the chart she had sketched, made some mental calculations and then looked ahead at the clouds. 'Yes, southwest.'

Pita immediately changed course to the southwest, but the storm moved faster than they could move to get out of the way. Lightning flashed under the approaching clouds.

The squall winds pounced just as Pita and Ani were donning their wet-weather gear. Dark clouds soon blocked the sun, leaving Pita to steer by the ocean's swells. Before long, wind-whipped waves tossed the waka around so violently that Pita could no longer feel the swells. It became impossible to sail in the storm's turbulent winds. In any event, having no sense of direction, sailing had become pointless.

The conditions were as bad as those they experienced the previous December. Two things made their situation worse. First, boats like Manaia with heavy keels usually right themselves after being capsized; the keel's weight pulling them upright again, just as Manaia's had. But proas, with no keel, typically remain upside down after capsizing. Second, this time no one knew they were out there, hundreds of kilometres from land. Ani and Pita knew no one would be looking for them, anywhere.

The sail, even furled, added to their risk of being blown over. Pita and Ani muscled the whole rig, sail, boom and mast, down to the deck. While Ani tied the rig down, Pita improvised a parachute-like sea anchor from the head of Manaia's mainsail and set it in the sea from the ama side of the waka. Ani threw herself into the whare as the storm continued to dump rain, and the waka was pitched around by waves and wind gusts.

Tūpuna Rock

As Pita returned to the open side of the whare, a flash of lightning revealed Ani coaxing him into the whare. He climbed in and wrapped his arms around her, as she wrapped hers around him. Their wet-weather gear, life jackets and harnesses made it a lumpy embrace, but it provided both the comfort and the reassurance they needed.

The improvised sea anchor successfully stabilized the waka's motion and kept the open side of the whare facing downwind. Pita and Ani lay awake for hours, deafened by thunderclaps and blinded by lightning flashes through the whare's sailcloth roof. The flashes of lightning recalled Manaia's cabin light, submerged in her overturned hull, sparkling and flashing in rippling water.

As the hours passed slowly in the darkness, the waka's fluid motion in the raucous sea gave Pita and Ani some confidence that he would ride out the storm in one piece and right side up. Morning's first light enabled Pita to see that the whole rig was still onboard. That, too, was reassuring, but the winds and waves continued to rage. Shortly after sunrise, the sun peeked through a gap in the clouds, providing a brief but unambiguous sense of direction – they were being driven westward.

By early morning, the worst of the storm had passed. Pita inspected the lashings securing the kiato to the hiwi and ama, as well as the lines temporarily making the mast and sail fast to the deck. Exhausted and reassured that everything was secure, Ani and Pita allowed themselves to sleep.

The sky was clearing when they awoke near mid-day. Ready to get underway, they rigged the mast and sail, and hauled in the sea anchor. Pita took the helm, preparing to resume the southerly course.

'Ani, the wind has shifted to the southeast.'

'How are you getting a bearing?'

'I can feel the swells again, they won't have changed since yesterday. When we have that simple side to side rocking, the wind is coming over the port bow.'

'Can you sail to the south in this wind?'

'It's marginal. We'd make better progress on a south-westerly reach.'

'I was awake a bit in the early morning. It looked like we were drifting at about one metre per second – three point six kilometres per hour. Give me a minute.' Ani pulled the notebook from her dry bag and drew on her hand-drawn chart. 'Yeah, go ahead and steer to the southwest. But we can't afford to go any further west than that, so no more storms, okay?'

'Okay, no more storms.'

That wish was granted. The sky continued to clear through the afternoon. By nightfall the sky was as clear as it had been on their first night at sea.

On the southwest heading, the motion of the waka in the swells was changed. Rather than a simple side to side rocking motion, the swells lifted the ama and the stern together before lifting the bow. The result was partly one of the waka rocking on a diagonal axis, but there was more complexity to the motion than that; there was a circular motion best observed at the masthead. Pita quickly had the sense of it, while Ani didn't believe she could feel it accurately enough to steer by. She hoped for continued clear skies.

On the midnight watch, Ani once again had the clear skies she wished for. She watched Marere-o-tonga finally remain above the horizon at its lowest point. It was only a tenth of a thumb above the horizon, but the height was measurable. They were a bit less than one-and-a-half thumbs of latitude north of Whangārei. A little more mental calculation told her that their estimate of three metres per second on a beam reach was a good one. She could not resist waking Pita and gave him a shake.

'My watch already?'

'No, sorry. Marere-o-tonga is at its lowest point and still above the horizon. I want you to see it.'

Pita climbed out of the whare and sat beside Ani looking over the port bow. The bright star was hovering over the horizon, just one of the thousands of stars in the night sky. Ani was prepared for a good-humoured comment about nerd navigators,

Tūpuna Rock

instead, she heard Pita say exactly what she was thinking, 'We've dreamt of seeing that for more than three months. Thanks for waking me.'

~ ~ ~

The next day was so perfect it could have been faulted as boring. A warm southeast wind continued at about 15 kilometres per hour. The waka glided along on a beam reach rocked by gentle swells from the east. A tāmure Ani hooked, when cleaned and grilled provided food for the day.

Pita sat with his floppy hat pulled down over his closed eyes and a broad smile, steering the waka by managing the angle of the sail to the wind and the masthead position. Suddenly he tipped his hat back and opened his eyes looking first at the angle of the sail and then at the hiwi and ama to visualize the waka's centre of lateral resistance. 'Ani, I think I understand why this mast position works.

'It's wrong to think of the sail's effort acting perpendicular to the hiwi.' He paged through Ani's notebook to find one of his earlier drawings. 'Do you see? This is the traditional Western way of considering the forces. But there's a simplifying assumption in Western naval architecture that the boat's centre of lateral resistance and the sail's centre of effort are both near the boat's centreline. That isn't true for proas.

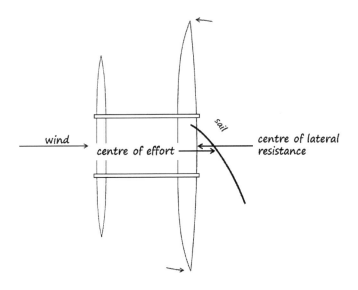

'The sail's effort must be more perpendicular to the sail.' He quickly made a new sketch with the resistance of the hiwi and ama drawn as a point between them, and the wind's effort acting perpendicular to the sail. 'I think it should look something like this.'

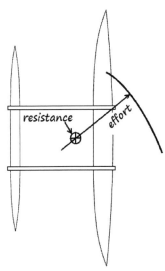

'This would be in balance, with the sail's effort pulling right through the centre of resistance. Do you see how much farther forward the sail is?'

'On conventional sailboats, with the sails and the centre of resistance both near the boat's centreline, how you draw the force wouldn't make much difference. But for a proa, with the sail above the hiwi and the centre of resistance somewhere out toward the ama, it matters a lot.

'You know, when I've seen pictures of proas out in Oceania, I've always thought they were carrying the sail way far forward, as if they were always running with the wind coming from astern. I see now they could have been smoking along on a reach. I should have known those Oceanic sailors really knew their stuff.

Before sunset, Pita trimmed the sail to a slight weather helm, so the waka would turn slowly toward the wind until the steering oar was dipped into the water, and handed the steering oar to Ani so he could inspect the waka. Hanging his head over the side of the deck, he studied the auhaka holding the two kiato to the hiwi. From the forward and aft ends of the deck he checked that all of the rigging was running freely. Moving to the ama side of the deck, he studied the aukaha securing the ama to each kiato. Along the length of the hiwi and ama, he checked that every flotation block was well secured. Finally, he sat down beside Ani.

'How does he look?'

'Everything that's supposed to be tight is tight, and everything that's supposed to run free is running free. I think he's good for another thousand kilometres.'

'We're nearing the end of our voyage. Wish we could just sail on for another week?'

'Ani, the fish you reel in are really great, honestly. But I'd love a bacon cheeseburger for a change.'

Ani laughed, 'We haven't had much variety in our diet, have we? If I never eat more fern root, that'll be just fine.'

'Enough joking. I haven't fulfilled my promise to Grandpa Koro. Everyone isn't safe yet. Still, this has been a beautiful day. I don't believe I've ever had a more beautiful day. One part of me would love to sail around for another week. But we know how nasty the weather can turn out here; we should use this patch of good weather to get back to Whangārei. And what did you say as we were leaving Tūpuna Rock? "We've been missing for too long".'

'Yes, we have been,' is all Ani could say in reply, still not prepared to think in depth about their family and friends suffering their apparent loss.

As Pita was climbing into the whare, leaving Ani on her evening watch, she asked, 'Do you want me to wake you see Marere-o-tonga at its lowest point?'

'Please.'

The moon was full, high in the sky behind them as Marere-o-tonga neared the horizon off their port bow. Ani roused Pita who emerged from the whare wrapped in the spinnaker quilt and sat beside Ani. As they watched Marere-o-tonga swing toward the horizon, Ani held out her thumb at arm's length at frequent intervals, finally announcing, 'That was it, eight tenths of a thumb.'

'It was one tenth last night. We made seven tenths of a thumb in a day.'

'That was seven tenths of north-south latitude, but we weren't going due south, we were going at a forty-five-degree angle to due south. Pythagoras says we have to multiply those seven tenths by the square root of two to figure how far we've gone today. That's just about exactly one thumb.'

'Two hundred and fifty kilometres.'

'Yup. We know our latitude, our speed and our course.' Ani did some quick mental math, 'We should be at the latitude of Whangārei before noon tomorrow.'

'Then we turn due west, and begin looking for islands and mountains we recognize.'

'If my estimate of our east-west longitude is close, we'll have one more night at sea before we see land. That would give us another chance to check our latitude.'

Tūpuna Rock

~ ~ ~

Pita took the helm midway through Ani's early morning watch, as the dawn's light began to obscure the stars. He was wide awake and eager to sail.

Shortly after sunrise, Ani stood up to stretch. Holding the mast strut for support, she faced eastward, soaking in the sun's warming rays for a few minutes. At last, intending only to have the sun warm her back, she turned toward the west-southwest. 'Pita!' she cried.

Pita looked up to see tears in his sister's eyes. 'Cloud tops,' she blurted, 'a line of them. Maybe they're just the tops of a long white cloud over the ocean, but maybe they're over a land a tūpuna navigator named for a long white cloud.'

Pita secured the mast position and stood beside Ani. He wrapped his arm around his big sister's back as she put her free arm across his shoulders. He, too, saw the line of cloud tops to the west-southwest. 'I believe our navigator has led us to the land called "Aotearoa".' He gave her back a squeeze before looking up into her misty eyes with an expression that said, 'Thank you,' in a way he could not verbalize.

Pita returned to tending the sail, while Ani remained standing, studying the horizon to the south-southwest. In time Pita looked up at Ani standing braced by one hand on the mast strut. Her sun bleached hair and shredded-sail piupiu were fluttering in the wind. Thanks to her mother's genes and more careful exposure to the sun, her shoulders, like her whole body, once burnt and peeling Canadian winter skin, had taken on the even hue of a lightly toasted marshmallow. 'Wow, sis. All of our canoe carving has put some muscles on your shoulders. Between your muscles and your math brain, you're going to be scaring the boys away.'

Ani gave a laugh, 'You don't think my piupiu and life-jacket fashion statement is flattering.'

'No.' He turned serious. 'I mean you're incredible. Really.'

'You're incredible too, Pita. Think about it. You aren't even old enough to drive a car, but you've skippered us most of the way back to Aotearoa, over nearly a thousand kilometres of open ocean. As for boys. If a guy is scared away by a few muscles and a brain, I'm not interested.'

Pita smiled, 'We'll never be quite the same, Ani. So many things will look different.'

Ani sat down on the deck opposite Pita and looked him squarely in the face. 'It's hard to imagine what will look the same.

'We've always loved Grandma Karani and Grandpa Koro. They are just such gentle, caring people.

'I paid attention when Karani took us, a couple of jet-lagged kids, to the beach to look at stars in in the dead of night. I never thought I would ever use the knowledge, but it was clear that the knowledge was sacred to this woman I loved, sacred to her iwi... our iwi. Without knowing why, I understood that the knowledge was sacred and committed it to memory. Now I understand why that knowledge is sacred – maybe more acutely than any other living Ngāti Te Taki-o-Autahi.'

'When Koro made a game of sailing for me, he didn't just teach this little guy about sail trim, although he did a heck of a job at that. He taught me to be a skipper, to keep everyone safe, to respect the boat, the crew. I learned that as a little boy.'

'You learned it well.

'I always thought of myself as a Canadian with a Māori mother. I love Canada, but I'm feeling more Māori now.'

'How could we not? The spirits of our Māori tūpuna saved us.'

'We're not saved yet,' Ani challenged with a wink.

'We have this waka; we know how to sail him; we know the way to Aotearoa, and our tūpuna are looking after us.'

~ ~ ~

'Do you hear that?' Ani asked.

'I was just beginning to think I heard something, a low droning sound.'

They scanned the horizon for nearly a full minute before a speck appeared over the southern horizon headed northward. As they watched, it became a plane on a course headed a few kilometres east of their position.

'We still have two flares,' Ani reminded. 'Should we launch a flare?'

'Flares are distress signals. Are we in distress?'

'We were a few months ago. But now, I'm enjoying the sail of my life. I'm sure not in distress.'

'Me too. So, no distress signal.'

'Think they'll see us anyway?'

They were seen. As it passed their position, three kilometres to their east, the plane began a gradual descent and a slow turn to the left. It returned at low speed one hundred metres above the sea and two hundred metres to their west.

Ani asked, 'Should we wave?'

'Don't do anything suggesting distress. Way out here in a tiny waka, they might interpret almost anything as a sign of distress. Just smile and make a "hi" sign. I'm going to make sailing along at six knots look as effortless as possible.'

'Right. I don't want anyone to come and "save" us now. We owe it to our tūpuna to sail all the way back to Aotearoa.'

The growling beat of the plane's propellers grew to a roar as it approached. Having not heard a mechanical sound for nearly four months, although not nearly as loud as the thunder two nights earlier, the roar was strangely unsettling for the voyagers.

It was a curious-looking plane for the twenty-first century, four propeller engines, straight wings and a strange boom extending from its tail. As each raised a hand in a motionless greeting, Pita and Ani could clearly see the kiwi bird in the roundel on the plane's side and crew members watching them through its windows.

'What is that?'

'Maritime patrol, a P-3 Orion. Canadian Forces have them, too. That thing extending from the tail is for detecting submarines.'

The Orion made a sweeping turn and returned for a low northbound pass to the east of the waka. Pita and Ani repeated their smile and motionless 'hi'-sign performance for the watching Royal New Zealand Air Force crew members.

After that second pass the Orion pulled up, returning to its original altitude and northward heading.

'If they thought we were in distress they would have stayed with us, circling until a ship or helicopter arrived to take us aboard.'

'Still,' Ani surmised, 'I think we'll be seeing more of New Zealand forces before we reach Aotearoa.'

'All we can do is keep sailing.'

'Toward Northland.'

'It is a beautiful day for sailing.'

11 Intercepted

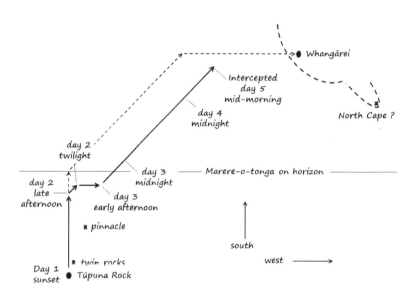

Four stories above the surface of the ocean, the softly-lit bridge of HMNZS Taupō provided a magnificent view of hundreds of square kilometres of open ocean. The crew on the bridge was quiet, all eyes scanning the horizon. At long last a sailor lowered his binoculars, 'There it is, sir, eleven o'clock, out near the horizon.'

Everyone on the bridge turned their gaze over Taupō's port bow. The commander picked up his binoculars and studied the unlikely sight, a small outrigger sailing canoe making way toward a coast two hundred and fifty kilometres away. 'I'll be… What are they doing out here?'

Taupō's navigator spoke up after a moment, 'They're on the same heading the Orion reported, headed toward Auckland.'

'They are now,' the commander allowed, 'but if they're traditional navigators, they may be planning to sail to somewhere in Northland.'

'Would they be this far out with no GPS?'

'Electronics or no, the big question is what a little waka ama is doing sailing out here at all. They didn't respond to calls from the Orion. Maybe they'll talk to fellow sailors.' The commander reached for his microphone, checked that the short-range VHS radio was set to the hailing channel and keyed the mic, 'Taupō, Taupō calling sailing canoe, sailing canoe.' There was no reply.

Each minute, as the distance closed between Taupō and the waka, the commander repeated the call, each time receiving no reply.

As they closed in, the commander called his chief engineer on the intercom, 'Chief, are you up for a look at a small boat?'

'Always, skipper.'

'We're closing in on that sailing canoe the Orion sighted. I'm going to send a boat out to investigate. I'd like you to go along to inspect her seaworthiness.'

'Can I sail it?'

The commander chuckled, 'That'll be up to her skipper.'

As Taupō approached the waka, her loudspeaker boomed, 'Sailing canoe, heave to. Sailing canoe, heave to.'

From their perch high above the water, the bridge crew watched a teenage girl in a white baseball cap and torn orange skirt furl the waka's sail and then look up at them. The boy manning the steering oar kept the oar in hand, looked up from under his floppy hat and waved.

Before long, one of Taupō's rigid hull inflatable boats was lowered and motoring out to the waka.

Tūpuna Rock

Taupō's commander watched as the boat tied up alongside the waka's hiwi. The chief shook hands with both of the waka's crew members and a long conversation ensued. The chief was known as a man of few words, so the commander was puzzled as he looked down on the extended conversation and finally picked up his mic, 'Chief?'

'Stand by, sir,' was the chief's immediate reply. The radio went silent again as the conversation between the chief and the waka's two crew members continued.

At long last the radio came alive, 'Skipper, you're not going to believe this.'

'Go ahead.'

'Remember the Canadian kids we were looking for back in December and January? It's them.'

The crew on the bridge exchanged dumbfounded glances. After his disbelief abated, the commander asked, 'Where have they been?'

'The navigator says they've been on some rock five-and-a-half degrees north and about seven degrees east of Whangārei.'

The commander looked to his navigator, 'What do you think, Katherine? The Kermadecs?'

'Must be, there isn't any other land to the northeast between here and Tonga,' she confirmed and then observed, 'Interesting navigation, latitude and longitude by Whangārei rather than the equator and Greenwich. There's a story there; I'd love to understand that.' She studied a chart plotter for a moment, 'It's actually a very accurate estimate for the Kermadecs.'

Back to the radio, the commander asked 'How did they get there?'

'They were blown there. They lost their sloop's rudder and were dismasted, but they kept it afloat. They finally beached the sloop on the rock.'

'Let's bring the survivors aboard. You stay with the waka and send them over in the boat. We'll arrange for another ship to hoist their canoe onboard or tow it to shore.' The commander could see the boy's floppy hat shake vigorously from side to side.

'Skipper, they don't want to leave their waka.'

Having the authority to remove shipwreck survivors from a makeshift craft in New Zealand patrolled waters, the commander thought for a moment, 'Chief, you stay on the waka with its skipper and send the navigator over in the boat.' The commander watched the chief negotiate with the waka crew.

'Skipper?'

'Go ahead.'

'One condition. The navigator doesn't want to be told our current position, or even compass directions.'

The commander saw a look of intense curiosity sweep over his own navigator's face. 'Okay, we'll cover our instruments.'

~ ~ ~

Having not stood on a stable surface for days, Ani's first steps on Taupō's deck were wobbly. She gained some sea legs for the big ship as she was led up to the bridge, but the scene inside the bridge was disorienting. Ani had not had a flat ceiling over her head since December and the only flat surface she had seen for months was the sea; suddenly she found herself surrounded by them. The interior of Taupō's bridge was an orderly arrangement of straight lines and flat surfaces, including the windows that wrapped in flat facets around the bridge forward and abeam. They were tinted and tilted downward, sheltering the bridge from the bright sun. She hesitated as she entered the bridge to find an outstretched hand there to steady and welcome her.

'Welcome aboard. I'm Jack Thomas, Taupō's commanding officer.'

'Thank you, I am Anna Dingstad,' Ani explained, her Canadian accent evident to the Kiwi crew.

There was a moment of silence as the casually-but-crisply-dressed bridge crew sized up the young woman they had brought up from the ocean. Sun-bleached blonde hair tumbled out from under the dirty white baseball cap on which they could decipher a Vancouver Whitecaps logo. The neck and arm holes of her life jacket revealed Māori designs on the bikini top she wore beneath it. The commander studied the high-visibility orange polyester her skirt was fashioned from. 'Storm jib?' he asked.

'It got so shredded it wasn't of any use as a sail anymore.'

'Can I offer you any clothes?' the navigator offered. 'You're about my size.'

'Ah, sorry,' the commander interrupted, 'I should have introduced our navigator, Katherine Gordon.'

'Nice to meet you. And thanks for the offer. But I've gotten comfortable in my voyaging rags.'

'We were looking for you,' the commander felt compelled to explain, 'but in the wrong area, I'm afraid. Your dinghy and rudder were found way to the south off the Coromandel, carried southward by the current. Our search effort focused on the current's track to the south.'

'I wondered about that. We tried at first to outrun the storm by motorsailing northward. We sent texts to our family explaining our plan, but I don't think we ever had cell reception. My brother started sending mayday calls as soon as we lost our rudder, but we weren't sure anyone received them. After we were rolled and dismasted, the winds and waves took us northward. We didn't know the direction at the time. The seas were so confused in the storm, and the clouds were so thick until the last day or two, that we had no sense of direction at all; we had lost our compasses and GPS when we were rolled. We had absolutely no idea where we were when we beached the boat. I don't expect it was possible for you to model which way the winds and waves had taken us.'

'You stayed afloat!' the commander noted with surprise. 'We thought losing the rudder would have left a big enough hole in your sloop to sink it pretty quickly.' He saw that Ani understood that the search had shifted to looking for them clinging to wreckage and finally for bodies and felt no need to continue.

'I think my brother nearly drowned down in the bilge plugging the hole the rudder stock left, but he got it plugged well enough, and we stayed afloat for another five days, until we beached the boat.'

The mention of drifting for so long with no navigation equipment or sense of direction piqued Katherine's curiosity. 'You told our chief that you'd been five-and-a-half degrees north and about seven degrees east of Whangārei, how did you figure that out?'

Ani explained at length measuring latitude in thumbs and building the observatory to find solar noon and estimate their longitude before her phone battery died.

Katherine was spellbound, 'So, since your phone battery died...'

'We had a solar-powered pocket calculator that was really helpful for designing the waka, but as far as navigation goes, it's just been the sky, the ocean, our brains and hands... ...and the spirits of our tūpuna.'

The commander was intrigued by the mention of tūpuna, 'Do Canadians call their ancestors, "tūpuna"?'

'We're Māori as well, our mum is Ngāti Te Taki-o-Autahi, from Whangārei.'

In the shade of the bridge, Ani had taken her sunglasses off, and, under the Vancouver Whitecaps cap and blonde hair, the commander saw Ani's brown eyes and oval Māori face.

'Our normal procedure is to sink an unseaworthy boat as a hazard to navigation,' the commander explained before adding quickly, 'I would not sink your waka. The alternative is to have a larger ship haul it ashore. We'll arrange that.'

'He's not unseaworthy. We've come more than eight hundred kilometres, and he's just fine. We're only another two or three hundred kilometres away from Northland.'

'You understand it hasn't been properly inspected...'

Ani interrupted, 'We have to sail our waka back to Aotearoa... to honour our tūpuna; their spirits taught us where we were and how to sail home.'

The commander's face softened, 'I understand. I have Māori blood as well, Te Whānau ā Apanui,' he added, naming his iwi, 'but our first responsibility is to get shipwreck survivors safely ashore.'

'We stopped being shipwreck survivors months ago. Then we became navigators and boatbuilders. Now, we are Māori voyagers.'

The commander called headquarters with the satellite phone on the speaker. 'Our survivors are of Māori ancestry and, out of respect for their ancestry, request that we allow them to sail their vessel back to Whangārei.'

'Stand by, Taupō,' crackled the reply, 'We've got to consider. We have already advised the Canadian Embassy that their citizens had been found alive. They are eager to have them ashore as quickly and safely as possible.'

Ani could not help interrupting, 'Begging your pardon, sir. This is Anna Dingstad. Please ask the Canadians what they would do if they found First Nation paddlers in a dugout canoe offshore headed toward Haida Gwaii.' She handed the microphone back to the commander.

'Thank you, ma'am, we'll talk to them. Can you repeat the paddlers' destination?'

'Stand by.' Katherine handed Ani a pen and pad of paper onto which she scribbled the words and handed it to the commander. 'That is "Haida Gwaii". I spell Haida: hotel, alpha, india, delta, alpha, Haida. I spell Gwaii: golf, whiskey, alpha, india, india. Gwaii. Haidi Gwaii.'

'Thank you. Meanwhile, Taupō, please assess the situation, the physical condition of the survivors, the seaworthiness of their vessel and so on... ...and make a recommendation for how to proceed.'

Ani's eyes followed the commander's as he looked out the port-side windows where the waka had been, then forward, then to starboard all to find the waka missing. He grabbed his VHF microphone. 'Chief, where are you?'

After a short delay, muffled laughter was heard on the radio. 'I'm just checking out the waka, sir. Per is teaching me how to sail it. Look to starboard. We'll be coming up from astern.' Ani joined the commander on the starboard side of the bridge as he watched the waka glide past. It was making five or six knots with the chief at the steering oar and Pita in his floppy hat trimming the sail.

'My brother is a natural-born sailor,' Ani boasted. 'The ocean talks to him through the hull – and his bony butt,' she added with a laugh. 'When clouds move in, and we can't see the sun or any stars, he can hold course just by feeling how the waka rolls in the swells. Even when it's clear, he prefers to steer by the swells. I check him when the sun or a star I know is near the horizon, and he's never off by more than a few degrees.'

The commander took in Ani's regard for her brother's skill and smiled.

'Would it be okay if I had our medic give you a quick check-up? The only medic onboard is male, so I'd have Katherine stay with you.'

'If your medic thinks I'm fit, can we sail our waka on to Aotearoa?'

'We've got a lot to consider. Your health is only one piece. But it's important.'

'I'd be happy to have a check-up. Haven't had one for months,' she added with a smile.

'Katherine,' the commander instructed his navigator, 'please take Miss Dingstad down to see doc. Stay with her as she is examined.'

'Sir,' Ani asked, 'on the rock, as we were learning from the spirits of our tūpuna, we began calling each other by our Māori names, "Ani" and "Pita". Please call me "Ani".'

'Āe, Ani,' the commander complied.

As Ani and Katherine were about to turn and leave the bridge, the commander observed, 'You call it a rock, but it's two kilometres across.'

Ani had to think about it. 'I guess it's because we've grown up in British Columbia. If it doesn't have thirty-metre-tall trees on it, it's a rock. What do you call it?'

'We call it "Macauley Island".'

Tūpuna Rock

'To us, it's "Tūpuna Rock".' No other name would ever seem right to her.

As Ani and Katherine were making their way out of the bridge and down the companionway, Katherine began, 'So tell me more about how you...'

The commander raised the chief on the VHF radio, 'So Chief, what you make of the waka?'

'Well, sir, some bits above the waterline look pretty crude. But below the waterline it looks like a race boat. The underwater lines are beautiful; I'm not surprised we tracked it making five knots in a modest wind. It's dug out from a driftwood log. It's solid and well-engineered. For flotation, they carved foam from a floating dock that washed up on the shore where they were shipwrecked and secured it with halyard. They saved every bit of line – halyard, sheet and anchor rode – from the sloop they lost. All the lashings are stout nylon or polyester, easily five or ten times stronger than needed.'

'Summary?'

'In stable weather, I believe this kid and his sister could sail this waka across the ditch, all the way to Oz.'

The commander laid out one of the paper charts kept as a backup to the electronics, Chart NZ 222, the Kermadecs. The chart was a metre long but covered only 300 kilometres of the vast South Pacific Ocean. Macauley Island, which lay in the centre of the chart, would have been covered by the width of a pencil lain on the chart. The commander studied the detail of the tiny island for a moment and then returned to the satellite phone to headquarters. 'I believe our survivors' story. But can you have intelligence examine satellite imagery of Macauley Island from January until last week. Look for a wrecked sloop in the cove on the southeast corner, signs of life on the northeastern edge, in the gullies, on the beach.'

'Will do.'

Looking at the chart again, the commander picked up a pen, struck out the name 'Macauley Island,' and carefully printed 'TŪPUNA ROCK.'

Ani and Katherine were still talking navigation when they returned to the bridge.

'So, Ani. If we let you sail onward,' the commander began, 'what is your sail plan?'

'I estimate that we're currently about one third of a degree north of the latitude of Whangārei.' The commander saw Katherine, standing behind Ani, nod in agreement. 'There are now clouds on the western horizon – long white clouds – Aotearoa. It's hard to know exactly how tall they are. My guess is they're five thousand metres tall, so I calculate we're about two hundred and fifty kilometres from Northland.' Katherine, still behind Ani, smiled and nodded again in agreement.

'So, our sail plan is to continue to the southwest to the latitude of Whangārei. The weather looks like I'll be able to get a good fix on our latitude at night using the star our Karani calls "Marere-o-tonga". Once on the latitude of Whangārei, we'll sail due west looking for islands and peaks we recognize. Our goal is to land the waka on the beach in Smuggler's Bay just south of Whangārei Heads. We know the beach; we've gone there often, ever since we were kids. It's a safe place to land.'

The commander took it all in and then simply nodded. He picked up the handset and called headquarters on the satellite phone, using the handset so Ani could not hear anything headquarters might say in reply. 'My engineer believes the waka is sound and fit for at least another thousand kilometres in all but the worst weather. They are wearing suitable PFDs (personal flotation devices – life jackets) and harnesses. They have more than a week's worth of food and water on board.

'My medic says their navigator, Miss Dingstad is in very good health. She is a skilled traditional navigator. She has a sound sail plan. There is no need to worry about them losing their way unless the weather closes in on them for an extended period.

Tūpuna Rock

'Mister Dingstad, has proven himself a highly capable skipper. We are going to bring him aboard so our medic can check him out.

'Assuming doc gives him a clean bill of health, here is my recommendation:

'We provide a radar reflector to hang on their mast and a handheld VHF radio. We provide flares and a torch for signalling. We offer a handheld GPS and hand-bearing compass, which I don't expect them to need or use.

'And then we release the survivors, to continue their voyage home to Aotearoa in their waka. Taupō will follow at a distance, maintaining radar contact until they are safely ashore. If the weather or anything threatens their safety, Taupō will promptly take the survivors onboard and take their waka in tow.'

The commander stopped transmitting and addressed Ani, 'I hope you can accept that.'

'Yes, thank you.'

The commander listened at length on his handset. He was being given a debrief on the satellite images of Macauley Island. They showed Manaia on the rocks in the cove in January and then washed away, the genoa stretched over the gully and the waka taking shape on the beach. Once they had pulled up the correct high-resolution images, the analysts at intelligence were heartbroken that their detailed search for the missing teens had been to the south, and they had not seen the signs of life so far north on Macauley months earlier. 'Please tell them it's understandable,' the commander asked and then wrapped up saying, 'Thank you, I will pass that along.'

The commander turned to Ani, 'Your parents are on their way back to Aotearoa New Zealand. I can only imagine how relieved they are.'

In the full embrace of the Royal New Zealand Navy, Ani had begun to emerge from survival mode and feel emotions from which she had been disconnected since December. 'These months must have been terrible for them,' she said as she teared up. Reading the commander's sense of propriety, Ani turned to her fellow navigator, who extended her arms toward her. Ani fell into those arms and wept silently.

After a quiet minute, the commander retrieved his life jacket and put it on. 'Let's send your brother in to see doc. Will you show me your waka?'

Ani perked up, 'Gladly.'

'Just a second,' Katherine reached into a locker and pulled out a scarf. 'Sometimes I wear this so my PFD doesn't chafe my neck. Please take it.'

Ani accepted the scarf, wrapping it around her neck and tucking it under the collar of her life jacket, 'Thanks so much.'

~ ~ ~

The commander and chief exchanged places. The commander to the waka and the chief to Taupō's boat. The commander took in the rāpaki Pita had fashioned from the sail bag. On the face of this accomplished sailor, whom he assumed had not shaved in months, or perhaps ever, he saw only the slightest wisp of a moustache, remembering the age of the sailor once believed lost was only fifteen. 'Your sister speaks highly of your sailing ability.'

'So do I,' announced the chief.

'My sister is one of the world's best non-instrument navigators, and a darned good hull designer. Together,' he smiled, 'we're a pretty good team.'

'I would like to have our medic examine you. Just to be sure you're fit to sail onward. Is that okay with you?'

'We have to sail the rest of the way. We owe it ...'

'Your sister explained. I understand.'

'Take me to your medic.'

Pita and Ani switched places, leaving the commander and Ani in the waka as the boat motored back toward Taupō.

'Would you like some smoked haku?' Ani offered as she unwrapped a piece from a patch of sailcloth.

The commander's first bite was small and tentative, but he then smiled and took a big bite, 'Wow, it's good!'

'It took us a few tries to get the recipe right, but yeah, it is pretty good.'

'You said you had plenty left?'

'Enough to get us to Australia. The fishing has been good, so we've been eating fresh fish most of the voyage down from Tūpuna Rock. As I said, we've still got two twenty-litre jugs of water left, there and there,' she explained pointing toward each end of the waka deck. 'It might not be enough to get to Australia, but it's a whole lot more than we need to get to Aotearoa – even if we do some washing up before we get there.'

It was a sound, well-provisioned boat with a highly skilled crew and a favourable weather forecast. The commander was already sold, but while waiting for Pita's return, Ani described sailing the waka: steering by dipping rather than twisting the steering oar, shunting, where to place the foot of the mast on different points of sail. The more she explained, the more questions he had, not challenges, but rather a desire to learn more about the technology that had carried their common ancestors to Aotearoa.

Facing away from his ship, the commander soaked in the experience of rocking gently in a tiny waka in the vast ocean. Noting the cloud tops peeking over the horizon to the west-southwest, he spoke, embellishing Ani's name, 'Ani Te Apārangi,'

'Te Apārangi?' Ani wondered aloud.

'You found your way to the long white cloud. The first person to find the long white cloud, to understand nature's signs and discover our country lying beyond the horizon was Hine Te Apārangi.'

'I thought it was Kupe.'

'Kupe was Hine Te Apārangi's husband. Kupe was a hell of an explorer and skipper,' the commander allowed before adding with a smile, 'blokes can navigate, too. But yeah, Aotearoa, the land of the long white cloud, was discovered and named by a woman.'

Continuing to study the horizon, the commander explained, 'Out here, Taupō's radars can't see land in any direction, but our tūpuna could see beyond the horizon. They taught us how, and how to find our home. Well done, Ani. Ka pai.'

~ ~ ~

As the returning boat approached the waka, Pita gave Ani a thumbs up, and then, raising a handheld radio, 'A VHF with a live battery, sis.'

Pita and the commander exchanged places, Pita to the waka alongside Ani and the commander to the boat with the chief. A radar reflector, flares and a signalling light were loaded onto the waka.

'Do you want a handheld GPS and a hand-bearing compass? I told headquarters I'd offer.'

'Thank you, but no,' Ani answered politely but firmly. 'We can find the way.'

'Fair winds, son,' the chief wished Pita.

'Thanks, mate.'

'Wait,' Ani called as the boat was about to shove off. The commander saw her offer five twigs and a bottle lashed together in a purposeful way. 'This is my solar compass. I don't need it anymore. You should have it onboard Taupō,' and then added with a smile, 'just in case all your electronics fail. I explained to Katherine how to use it.'

The commander studied the gift. 'Thanks. We'll keep it onboard Taupō. All right then. Haere rā,' the commander offered as the boat moved away from the waka.

'Haere rā,' the voyagers called back as the boat turned toward Taupō.

~ ~ ~

Tūpuna Rock

Taupō, lingering on the horizon, was soon far behind the little waka. When Ani reckoned they had reached the latitude of Whangārei, she commanded the turn to due west. As the sounds of the wind and water sang to the silent crew, the mast was moved and the sail trimmed for the broad reach.

~ ~ ~

The handheld radio crackled, disrupting the happy gurgling sound of the waka making way, 'Taupō, Taupō calling Tūpuna Rock, Tūpuna Rock, this is Taupō, Taupō. Over.'

Pita looked at Ani, 'We never christened our waka. But they must be calling us.'

'It's a good name, "Tūpuna Rock". Let's claim it.'

Pita picked up the radio, 'Tūpuna Rock calling Taupō. Over.'

'Tūpuna Rock, switch to two-one-alpha. Over.'

Being asked to switch to a non-emergency channel was a good sign. Pita switched the radio to channel 21A. 'Taupō, this is Tūpuna Rock. Over.'

The commander's voice came through, 'Tūpuna Rock, headquarters has arranged for another vessel to continue your escort. That vessel monitors channel one-six for emergency calls but remains silent otherwise. We will watch your radar signatures until we are confident you have visual contact; then we will break off and get back to work. Ka kite anō tāua i a taua, Tūpuna Rock. Taupō out.'

Neither Pita nor Ani recognized the commander's Māori goodbye indicating that he expected to meet them again and replied with a simple goodbye. 'Hei konā rā, Taupō. Tūpuna Rock out.' Pita switched the radio back to monitor channel 16.

The gurgling sound of their waka making way regained its dominance. Tūpuna Rock sailed on for more than an hour with neither Ani or Pita seeing anything on the horizon except Taupō's bridge and radar dome to the east. Finally, all signs of Taupō disappeared over the horizon, and they became aware of something on the horizon to the south-southwest.

Intensely curious about the silent vessel the Navy had sent to escort them, Pita and Ani watched as a smudge on the horizon morphed into two smudges and, at length, into two sails. At long last under traditional Polynesian sails like the rā kautu they had fashioned themselves for their waka, they could make out a traditional double-hulled voyaging canoe, its crew on deck watching them.

Watching their new escort approach from what she knew to be south-southwest, Ani advised Pita, 'They're probably coming from Tāmaki Makaurau. I'm confident of our course. Maintain this heading.'

'Yes, ma'am. I'm a believer.'

~ ~ ~

Ani furled their sail as the voyaging canoe, Hinemoana, swung around and pulled alongside Tūpuna Rock.

'Kia ora, e hoa mā. Ko Hoturoa taku ingoa.' Hinemoana's skipper, Hoturoa, greeted Pita and Ani and introduced himself in Māori.

'Kia ora, Hoturoa.'

'We got a call from the Navy. They told us about your voyage. They are very impressed. We're all very impressed. They didn't think it was right for a four-hundred-tonne stinkpot to escort you the rest of the way to Aotearoa and asked if we could gather a crew and come out to meet you. We thought we needed some kaumatua on the crew so we picked up these two.' At that point Hoturoa motioned toward the voyaging canoe's whare, from which Karani and Koro emerged.

Pita and Ani were speechless, staring into their grandparents' eyes in disbelief. Tears welled up in everyone's eyes.

Karani and Koro studied their grandchildren. Pita's hair, short when they last saw him, had grown out into an unruly mop. Koro noted a scar on Pita's jaw but said nothing of it. Though clad in scraps of sailcloth and a sail bag, they looked fit, tanned and athletic.

Tūpuna Rock

Ani was the first to speak, 'Karani, it was you who taught us how to find home, taku kāinga.'

Karani gave a puzzled look.

'Remember standing on the beach holding your thumb under Māhutonga on winter mornings and telling us about Marere-o-tonga?'

'You remembered?'

'We remembered. It was what we needed to know to understand how far north of your beach we were, and begin to find our way back. We figured out the rest.'

Hoturoa studied Karani and Koro and then Ani and Pita, in proud awe of the Māori teachers and young voyagers. 'The rest' was a great deal to figure out.

Koro caught Pita's eyes, glanced toward Ani and then returned to Pita with a smile, 'You kept everyone safe.'

Pita returned the smile. 'I could never forget. But I lost Manaia; I'm sorry.'

'No worries.' Koro looked at the waka's mast, yard and sail. 'You brought some of her back. Anyway,' he added looking down at the waka's hiwi, 'I like this waka better. Ka pai.'

'Pay no mind to where we came from,' Hoturoa instructed.

'Don't worry, she already told me not to,' Pita nodded toward Ani.

'We'll follow at a distance.' Hoturoa explained and then asked simply, 'Lead us to Aotearoa.'

'Northland, Whangārei,' Ani refined their destination.

'Āe, ki Te Tai Tokerau,' Hoturoa confirmed their destination in Māori.

Pita let out the sail and they led onward on a broad reach.

After the sun set behind the long white cloud, Pita took in the dark sea ahead and bright stars that surrounded them. 'Navigator, I need to get a solid night's sleep,' he announced as he handed off the steering oar to Ani, before climbing into the whare and wrapping up in the spinnaker quilt. 'We've got some coastal pilotage ahead of us tomorrow, manoeuvring our waka around islands and into harbour. I'll really have to earn my keep as a helmsman. 'Til then, it's all up to you.'

Ani steered into the night by the moon and stars as Hinemoana followed at a distance of nearly a kilometre. From different waka, Ani and Karani watched Māhutonga and Marere-o-tonga, wheel slowly around the celestial pole. They could not see each other across the darkness, but, when Marere-o-tonga passed closest to the horizon, granddaughter and grandmother both held out their open hands on outstretched arms and measured its height above the moonlit sea with their thumbs. In the first light of morning, they would see the familiar islands and coastal peaks of Northland.

Tūpuna Rock

Klaus Brauer

Tūpuna Rock

Author's note

We are all indebted to a few traditional navigators who chose to share their knowledge with individuals from across the ocean before the art and science of Pacific wayfinding were lost to humanity. The navigators Basil Tevake, Hipour and Mau Piailug will be remembered for lighting the torches of knowledge others would carry. These navigators have now all passed away. Thankfully, those they taught have documented and practiced the knowledge and, in turn, taught countless others. Thanks to the work of David Lewis, Nainoa Thompson, Sir Hekenukumai Busby and too many others to name, it is now difficult to imagine that the knowledge can ever be lost.

However, knowledge can be preserved while its originating culture is crushed. Oceanic cultures have been under siege ever since their 'discovery' by European explorers. The indigenous people of Oceania – Polynesians, Micronesians and Melanesians – have suffered many of the same indignities suffered by indigenous peoples throughout the world, more often than not becoming a disrespected underclass on their own islands.

The purpose of is this book, then, is not to expand on the art and science of wayfinding. Far beyond teaching some practical science and mathematics, its purpose is to bring an essence of Oceanic wayfinding and seafaring to a wider audience with the aim of building pride and respect, if not awe, for the genius of the people of Oceania.

Klaus Brauer

Acknowledgements

Heidi, my spouse, partner and Viking muse must be acknowledged before all others. Without her love, patience and support, this book (and most good things in my life) would never have taken shape. She is also always my first, last and most helpful editor.

With apologies to parents who are not similarly fortunate, I must thank our children who have different interests and strengths. The relief from competition resulting from their different interests, and compatibility resulting from their appreciation of each other's strengths, provided a model for the mutually supportive sibling interaction between Ani and Pita in this book.

In addition to his vast expertise in sailing canoe design, I am grateful to Gary Dierking for his early editorial advice.

Louise Russell provided many helpful inputs which led me to make this a richer book.

Tiraroa Reweti was most helpful in nudging me to reshape my characters and to avoid some cultural faux pas.

Thanks are due to the Media Centre of the Royal New Zealand Navy for its assistance in researching details of HMNZS Taupō.

Heidi and I are inspired by Jolene Busby's continuation of her grandfather Hekenukumai's legacy, particularly by her dedication to the children – the tamariki.

Hoturoa Barklay-Kerr, CNZM, has made Heidi and me feel at home in the voyaging whanau. We are most grateful for the embrace of Hotu and the entire Te Toki Waka Hourua crew. As with Jolene Busby, we are inspired by their dedication to the children of Aotearoa New Zealand and of the greater Pacific.

Tūpuna Rock

Glossary

āe – yes
āhuruhuru – goatfish
ama – outrigger
Aotearoa – New Zealand, literally 'land of the long white cloud'
aukaha – lashing
e hoa mā – mates
e noho rā – 'good bye' said to someone who is staying
haere rā – 'good bye' said to someone who is leaving
haku – kingfish, a popular game fish
hei konā rā – 'good bye' said on the telephone or to someone who is staying
Hinemoana – 'Woman of the sea', the name of a replica of a traditional Polynesian voyaging canoe.
hiwi – the main hull of an outrigger canoe
ingoa – name
iwi – tribe
kāinga – home
hikurere – shawl
ka kite anō tāua i a tāua – 'good bye' said to someone one anticipates seeing again
ka pai – It's good
karani – granny, grandmother
kaumatua – elders
kia ora – hello, literally 'good health'
koro – granddad

Klaus Brauer

Motu Muka – Lady Alice Island, the westernmost large island of the northern Marotere / 'Hen and Chickens' islands
Ngāti Te Taki-o-Autahi – People of the Southern Cross, the name of a fictional tribe, or 'iwi' from Northland, the northernmost region of Aotearoa
piupiu – a skirt made of flax
proa – (not Māori, from Malay) an outrigger sailing canoe that keeps its outrigger to windward when sailing
rā kautu – a traditional Polynesian sail with a vertical forward edge (luff) and slanted aft edge (foot)
rāpaki – kilt or skirt
taku – my
tamariki - children
Tāmaki Makaurau – Auckland
tāmure – snapper
tohunga – expert, master
tūpuna – ancestor, ancestors
uff da – (Norwegian esp. American diaspora) an interjection indicating something is unpleasant or uncomfortable
wāhine – woman/women
waka – canoe (waka ama – outrigger canoe)
Whangārei – (pronounced 'fah-ngah-rey') the largest city in Northland, the northernmost region of Aotearoa
whare – (pronounced 'fah-ray') house, the deck house on a voyaging canoe
Whatupuke – (pronounced 'fah-too-poo-kay') the middle large island among the northern Marotere / 'Hen and Chickens' islands

Tūpuna Rock

Star Names

Māori and common names for stars used by Ani and Pita:

'anchor shackle' – Rehua / Antares [rises 121° 04']
'below Orion' – Taumata-kuku / Aldebaran [sets 289° 15']
Māhutonga – Te Taki-o-Autahi / Gacrux [declination -57°13.6'] (Ani and Pita refer to this star by the name of its constellation.)
Marere-o-tonga – Marere-o-tonga / Achernar [declination -57°08.3']
'midnight rise' – Poutūterangi / Altair [rises 79° 40']
'midnight set' – Puanga-hori / Procyon [sets 275° 59']
'north of anchor' – Ruawāhia / Arcturus [rises 67° 46']
Orion's belt – Tautoru / Alnilam [sets 268° 37']
'past midnight' – Whānui / Vega [rises 43° 31']
Sirius – Takurua / Sirius [sets 250° 31']
'sunset star' – Whiti-kaupeka / Spica [rises 103° 04']
'Tau Hou' (New Year) – Te Rā o Tōnga / Atria [declination -69°03.5']

Rising and setting bearings are as viewed from 'Tūpuna Rock' at 30° 14' South latitude.

Klaus Brauer

Celestial latitude at the time of the migration

The author is not aware of any record of precisely how the navigators of the Polynesian migration to Aotearoa determined latitude. However, as one who has viewed the heavens with the eye of a celestial navigator for more than fifty years, he is certain that navigators of their skill and sensitivity to the sky and sea were keenly aware of 'circumpolar' stars that, at their lowest point, passed close to the horizon at the latitude of their intended destination. 'Karani', a woman with a true Māori sensitivity to the natural environment, would have been similarly aware.

As a result of the precession of the Earth's axis over the centuries since the Polynesian migration to Aotearoa, the stars Karani used to determine the latitude of Whangārei (Gacrux and Archernar) would not have been as useful to the ancient navigators as they are today. As it happens, at the time of the migration, Acrux (the star opposite Gacrux in Māhutonga, the Southern Cross) and Peacock would each have stood approximately five degrees (2.2 thumbs) above the horizon at their low points when viewed from Whangārei. Like Gacrux and Archernar in the present day, Acrux and Peacock would have provided the ancient navigators excellent latitude references for Whangārei or any landfall in Northland.

If an oral tradition of the ancient navigators using such circumpolar stars to determine latitude did, indeed, survive, it seems likely that the identities of the useful stars would have evolved in the retelling to match the heavens above. There has been uncertainty with regard to the traditional Māori names for many stars, the evolution of oral traditions to match the evolving heavens may explain some of that.

Bibliography

Allen, Jennifer. *Mālama Honua*. Ventura, California, USA: Patagonia Books, 2017.

Bader, Hans-Dieter & Peter McCurdy, Eds. *Proceedings of the Waka Moana Symposium*. Auckland, New Zealand: New Zealand National Maritime Museum Te Huiteananui-a-Tangaroa, *1996*, 1999.

Best, Elsdon. *The Māori Canoe*. Wellington, New Zealand: A. R. Shearer, Government Printer, 1976 (first published 1925).

Best, Elsdon. *The Astronomical Knowledge of the Māori*. Wellington, New Zealand: V.R. Ward, Government Printer, 1986 (first published 1922).

Colenso, William. *On the Vegetable Food of the Ancient New Zealanders Before Cook's Visit*, Transactions and Proceedings of the Royal Society of New Zealand, Volume 13, Art. I., 1880.

Dierking, Gary. *Building Outrigger Sailing Canoes*. Camden, Maine, USA: International Marine / McGraw-Hill, 2008.

Dodd, Edward. *Polynesian Seafaring*. New York, USA: Dodd, Mead & Company, 1972.

Evans, Jeff. *Heke-Nuku-Mai-Ngā-Iwi Busby – Not Here by Chance*. Wellington, Aotearoa New Zealand: Huia Publishers, 2015.

Haddon, A. C. & James Hornell. *Canoes of Oceania*, Honolulu, Hawaii, USA: Bishop Museum Press, 2017 (first published in three volumes in 1936, 1937 and 1938).

Howe, K. R. Ed. *Vaka Moana – Voyages of the Ancestors*. Auckland, New Zealand: David Bateman, 2006.

Lewis, David. *We, the Navigators – The Ancient Art of Landfinding in the Pacific*. Honolulu, Hawaii, USA: University of Hawai'i Press, 1972, 1994.

Low, Sam. *Hawaiki Rising*. Waipahu, Hawaii, USA: Island Heritage Publishing, 2014.

Nelson, Anne. *Ngā Waka Māori – Māori Canoes.* Auckland, New Zealand: the Macmillan Company of New Zealand Ltd., 1991.

Norton, Elsie K. *Crusoes of Sunday Island.* New York, USA: W. W. Norton & Co Inc, 1958.

Thomas, Stephen D. *The Last Navigator.* New York, USA: H. Holt, 1987.